PEER-LED TEAM LEARNING

ORGANIC CHEMISTRY

PEER-LED TEAM LEARNING

ORGANIC CHEMISTRY

SECOND EDITION

Jack A. Kampmeier
University of Rochester

Pratibha Varma-Nelson
Northeastern Illinois University

Carl C. Wamser
Portland State University

Donald K. Wedegaertner
University of the Pacific

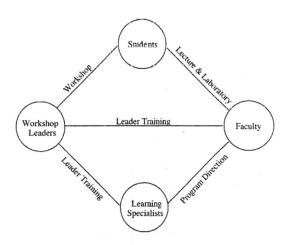

 PRENTICE HALL SERIES IN EDUCATIONAL INNOVATION

Upper Saddle River, New Jersey 07458

Project Manager: *Kristen Kaiser*
Production Editor: *Shari Toron*
Executive Marketing Manager: *Steve Sartori*
Executive Managing Editor: *Kathleen Schiaparelli*
Manufacturing Buyer: *Alan Fischer*
Copy Editor: *Brian I. Baker*
Art Director: *Jayne Conte*
Manager, Cover Visual Research & Permissions: *Karen Sanatar*
Cover Design: *Bruce Kenselaar*
Cover Images: *top*: Getty Images, Inc.; *bottom right*: Gary Buss / Taxi / Getty Images, Inc.;
bottom left: Nick Rowe / Photodisc Green / Getty Images, Inc.; *bottom center*: Getty Images, Inc.;
bottom right: Gary Buss / Taxi / Getty Images, Inc.

© 2006, 2001 by Pearson Education, Inc.
Pearson Prentice Hall
Pearson Education, Inc.
Upper Saddle River, NJ 07458

Pearson Prentice Hall™ is a trademark of Pearson Education, Inc.

Printed in the United States of America
10 9 8 7 6 5 4 3 2 1

ISBN 0-13-1855107

Pearson Education LTD., *London*
Pearson Education Australia PTY. Limited, *Sydney*
Pearson Education Singapore, Pte. Ltd.
Pearson Education North Asia Ltd, *Hong Kong*
Pearson Education Canada, Ltd., *Toronto*
Pearson Educacion de Mexico, S.A. de C.V.
Pearson Education—Japan, *Tokyo*
Pearson Education Malaysia, Pte. Ltd.

Preface

Introduction to Workshops for Organic Chemistry

Our early research and development convinced us that the Peer-Led Team Learning Workshop helped students acquire the habits of study and thought required to be successful in organic chemistry. In the first edition of this book, we collected successful Workshop problems from a variety of sources. Our purpose was to make these problems freely available to faculty in order to lower the activation energy required to implement the PLTL Workshop in their courses. Prentice Hall was a remarkable ally, publishing and distributing the text to faculty without charge. Since that time, we have learned more about constructing Workshops that help students build their conceptual understanding of organic chemistry.

This second edition is designed as a student workbook, to be used as the central focus of activity in a PLTL Workshop in organic chemistry. It is not a drill book, nor is it a self-contained guided-inquiry book. As with the Workshops themselves, this book is intended to be a companion to a textbook in a lecture course. The Workshop problems are challenging; they represent our sense of what we want our students to know, understand, and be able to do. Students need to prepare for these Workshop problems by studying the text and the lectures and by working the end-of-chapter problems ahead of Workshop time.

Our thinking about the book has been stimulated by the idea of cognitive apprenticeship. We came to the understanding that the structure of a problem could provide the cognitive modeling and scaffolding that students need to move to new levels of skill and insight. We also constructed the problems to encourage the discussion and debate among peers that is essential in building conviction and confidence. We know that the majority of the students in an introductory organic course will not make subsequent use or even retain most of the specific content of the course. But we also know that the BIG IDEAS and the new thinking skills that come from a yearlong study of organic chemistry are essential ingredients of a liberal education. For most students, these ideas and skills will be the long-term consequence of their study of organic chemistry. Accordingly, we have tried to make these outcomes explicit.

We have organized this book for intellectual, rather than temporal, coherence. The four of us use different textbooks to teach organic chemistry in different sequences. Nevertheless, we have found it easy to rearrange the Workshops to fit our own sense of timing. In most Workshops, we have given more than two hours worth of problems, recognizing that faculty will pick and choose those which serve their syllabus best.

We have many essential collaborators to thank. First among those are the hundreds of Workshop leaders and thousands of students who have helped to shape our ideas and the problems we pose. John Challice, Kristen Kaiser, and Prentice Hall have nourished this project from the start. Arlene Bristol and Korrie Sherry performed wonders, with patience and skill, to produce an attractive and readable book. This work was supported by NSF/DUE awards, 9450627 and 9455920.

Jack Kampmeier, *University of Rochester*
Pratibha Varma-Nelson, *Northeastern Illinois University*
Carl Wamser, *Portland State University*
Donald Wedegaertner, *University of the Pacific*

To the Student

An Introduction to the Peer-Led Team Learning Workshop

The Peer-Led Team Learning Workshop is a unique curricular structure that provides a weekly opportunity for you to engage with your fellow students in the process of constructing your understanding of organic chemistry and developing new problem-solving skills. As part of the process, you will talk, debate, discuss, argue, evaluate, ask questions, answer questions, explain your ideas, listen to the ideas of your colleagues, and, ultimately, negotiate your understanding of organic chemistry with them. This process of active, personal engagement with experimental observations and with the ideas of colleagues is the way that scientists construct meaning and understanding. Practicing scientists do this in a structure called the research group meeting. The Peer-Led Workshop is a research group meeting for undergraduates.

Each week's Workshop is built around a set of problems that are designed to help you probe and build your conceptual understanding of a BIG IDEA in organic chemistry. Simultaneously, the problems require you to learn and practice new problem-solving skills. The problems are challenging, so you will need the resources and imagination of the group to come to a good solution. We do not provide answers to the problems, because the point of the Workshop (indeed, the point of higher education) is to help you learn to construct, evaluate, and develop confidence in your own answers and conclusions. Ultimately, your understanding is personal and belongs to you. However, the process of testing and refining your understanding is dramatically facilitated by interaction with your peers.

Because interaction with your peers is the key to successful learning, each Workshop is guided by a Workshop leader. This leader is not a teacher, or a teaching assistant, or an expert authoritarian answer giver about organic chemistry. Rather, the leader is a fellow student-a peer-who has successfully completed the course and is chosen and trained by your teacher to lead your Workshop. The job of the leader is to guide the interactions among the disparate participants so that the group becomes a team in which the members work together to ensure that everyone learns organic chemistry. The leader's job is to make sure that everyone participates and that everyone learns. Your leader will be a role model, mentor, cheerleader, and friend.

Workshop problems are not homework; they are designed for cooperative teamwork during the Workshop. That does not mean that there is no homework in the course. You need to prepare for the Workshop in order to be a participating, contributing member of the Workshop team. This means that you have to study the text, attend and listen to the lectures, and do the assigned homework problems from the text before you come to Workshop. Each Workshop begins with a statement of **Expectations** that should help to guide your preparation for the Workshop.

You may be familiar with workbooks that drill you on the empirical knowledge associated with learning a particular subject. This is not that kind of workbook; we assume that you are already skilled at acquiring knowledge. Rather, the book is focused on the development of conceptual understanding and problem-solving skills-that is, on the organization and application of knowledge and ideas in organic chemistry. The book is a workbook in the sense that it provides space for you to make notes, record the results of the teamwork, and reflect on the point of the problem and the BIG IDEAS it involves. Each Workshop ends with a **Reflection** section that invites you to identify gains in your understanding and skill. If you use your workbook well, it should become an effective tool for review for exams.

There are many rewards to be earned from the study of organic chemistry. It is a powerful and wonderful way to understand numerous aspects of our natural world. The interplay of molecular structure and chemical and physical properties is one of the foundational ideas of modern science. In contrast to the prevailing student mythology, organic chemistry is a coherent, rational subject with an accessible theory that makes sense of a wealth of observations. It is also a subject of enormous practical consequence for our health and well-being. Furthermore, organic chemistry is exceptionally well suited to the development of powerful cognitive skills that are general and transferable to other areas of study and decision making. In 1964, Benjamin Bloom compiled a taxonomy of cognitive skills, organized in a hierarchy from the simplest to the most complex: knowledge, comprehension, application, analysis, synthesis, and evaluation. Successful work in organic chemistry requires you to function at all of these levels. The Workshop problems are designed to help you learn the requisite thought processes.

Finally, it is important for you to understand the central position of students-you and others in the Workshops. We cannot teach you organic chemistry; at best, we can help you learn the subject. In the end, you will not acquire our understanding. Instead, there is the exhilarating prospect that you will build your own view of organic chemistry and make your own contributions to our understanding. Thousands of students have already done that. We are grateful to them. It is a special pleasure to acknowledge the inspiring and motivating contributions of hundreds of Workshop leaders. They have tested most of these Workshops and provided a unique window on the student perspective.

Jack Kampmeier, *University of Rochester*
Pratibha Varma-Nelson, *Northeastern Illinois University*
Carl Wamser, *Portland State University*
Donald Wedegaertner, *University of the Pacific*

Periodic Table of the Elements

Main groups | Transition metals | Main groups

	1A 1 [a]	2A 2	3B 3	4B 4	5B 5	6B 6	7B 7	8B 8	8B 9	8B 10	1B 11	2B 12	3A 13	4A 14	5A 15	6A 16	7A 17	8A 18
1	1 H 1.00794																	2 He 4.002602
2	3 Li 6.941	4 Be 9.012182											5 B 10.811	6 C 12.0107	7 N 14.0067	8 O 15.9994	9 F 18.998403	10 Ne 20.1797
3	11 Na 22.989770	12 Mg 24.3050											13 Al 26.981538	14 Si 28.0855	15 P 30.973761	16 S 32.065	17 Cl 35.453	18 Ar 39.948
4	19 K 39.0983	20 Ca 40.078	21 Sc 44.955910	22 Ti 47.867	23 V 50.9415	24 Cr 51.9961	25 Mn 54.938049	26 Fe 55.345	27 Co 58.933200	28 Ni 58.6934	29 Cu 63.546	30 Zn 65.39	31 Ga 69.723	32 Ge 72.64	33 As 74.92160	34 Se 78.96	35 Br 79.904	36 Kr 83.80
5	37 Rb 85.4678	38 Sr 87.62	39 Y 88.90585	40 Zr 91.224	41 Nb 92.90638	42 Mo 95.94	43 Tc [98]	44 Ru 101.07	45 Rh 102.90550	46 Pd 106.42	47 Ag 107.8682	48 Cd 112.411	49 In 114.818	50 Sn 118.710	51 Sb 121.760	52 Te 127.60	53 I 126.90447	54 Xe 131.293
6	55 Cs 132.90545	56 Ba 137.327	71 Lu 174.967	72 Hf 178.49	73 Ta 180.9479	74 W 183.84	75 Re 186.207	76 Os 190.23	77 Ir 192.217	78 Pt 195.078	79 Au 196.96655	80 Hg 200.59	81 Tl 204.3833	82 Pb 207.2	83 Bi 208.98038	84 Po [208.98]	85 At [209.99]	86 Rn [222.02]
7	87 Fr [223.02]	88 Ra [226.03]	103 Lr [262.11]	104 Rf [261.11]	105 Db [262.11]	106 Sg [266.12]	107 Bh [264.12]	108 Hs [269.13]	109 Mt [268.14]	110 [271.15]	111 [272.15]	112 [277]	113 [284]	114 [289]	115 [288]	116 [292]		

*Lanthanide series

57 *La 138.9055	58 Ce 140.116	59 Pr 140.90765	60 Nd 144.24	61 Pm [145]	62 Sm 150.36	63 Eu 151.964	64 Gd 157.25	65 Tb 158.92534	66 Dy 162.50	67 Ho 164.93032	68 Er 167.259	69 Tm 168.93421	70 Yb 173.04

†Actinide series

89 †Ac [227.03]	90 Th 232.0381	91 Pa 231.03588	92 U 238.02891	93 Np [237.05]	94 Pu [244.06]	95 Am [243.06]	96 Cm [247.07]	97 Bk [247.07]	98 Cf [251.08]	99 Es [252.08]	100 Fm [257.10]	101 Md [258.10]	102 No [259.10]

[a]The labels on top (1A, 2A, etc.) are common American usage. The labels below these (1, 2, etc.) are those recommended by the International Union of Pure and Applied Chemistry.

The names and symbols for elements 110 and above have not yet been decided.

Atomic weights in brackets are the masses of the longest-lived or most important isotope of radioactive elements.

Further information is available at http://www.webelements.com

The production of element 116 was reported in May 1999 by scientists at Lawrence Berkeley National Laboratory.

Table of Contents

Workshop 1   Structure: Functional Groups   1

Workshop 2   Structure: Molecular Geometry and Bonding   5

Workshop 3   Structure and Properties   13

Workshop 4   Structure and Properties: Acids and Bases   22

Workshop 5   Reaction Mechanisms   30

Workshop 6   Stereochemistry of Alkanes and Cycloalkanes   37

Workshop 7   Alkenes: Electrophilic Addition Mechanism: Carbocations   42

Workshop 8   Alkenes: Reactions   49

Workshop 9   Free-Radical Reactions: Thermochemistry   57

Workshop 10   Organic Synthesis   64

Workshop 11   Chirality   69

Workshop 12   Nucleophilic Substitution Reactions   77

Workshop 13   Elimination Reactions   84

Workshop 14   Alkyl Halides and Alcohols: Synthesis   91

Workshop 15   Epoxides and Ethers   99

Workshop 16   Conjugated Systems   107

REVIEW   116

Workshop 17   Aromaticity   138

Workshop 18   Aromatic Substitution   144

Workshop 19   Pericyclic Reactions   153

Workshop 20   Aldehydes and Ketones   165

Workshop 21   Enols and Enolate Ions   173

Workshop 22   Ester and β-Dicarbonyl Enolates   181

Workshop 23   Carbohydrates   188

Workshop 24   Phenols   196

Workshop 25   Carboxylic Acids   203

Workshop 26   Carboxylic Acid Derivatives: Nucleophilic Acyl Substitutions   210

Workshop 27   Lipids   218

Workshop 28   Amines   227

Workshop 29   Amino Acids and Peptides   233

Workshop 30   Metabolism   241

REVIEW   247

WORKSHOP 1

Structure: Functional Groups

Purpose: "Function follows form" is a fundamental principle that helps us make sense of the physical and chemical properties (function) of molecules. There is little room for ambiguity about the structure (form) of a diatomic molecule AB: The two atoms must be connected to each other. With three atoms, A_2B, there are two different connectivities. Since carbon bonds to itself and other atoms, there are many ways to connect larger sets of atoms. This Workshop provides opportunities to explore the structures of constitutional isomers and investigate logical ways to find those structures. Remarkably, the structures of different isomers can be deduced from experimental observations of their properties. The logic also works the other way around: properties can be predicted from molecular structure. This Workshop begins to explore these fundamental patterns of reasoning; we will use them throughout the course.

Expectations: To prepare for this Workshop, you should review and understand the following terms and ideas: molecular formula; constitutional isomer; equivalent atoms and molecules; functional group structure and nomenclature; Kekulé structures; primary, secondary, tertiary, and quaternary nomenclature; and simple rules for naming organic compounds.

1.   The following isomers of molecular formula C_5H_{12} were treated under appropriate conditions to give all possible monochloro products, $C_5H_{11}Cl$, in which H's are replaced by Cl. These $C_5H_{11}Cl$ constitutional isomers, derived from each C_5H_{12} compound, could be separated either by careful fractional distillation or by gas chromatography (GC). Work together to find the structures of the following C_5H_{12} compounds and the corresponding $C_5H_{11}Cl$ derivatives.

   a.   A C_5H_{12} compound that gives three $C_5H_{11}Cl$ derivatives

   b.   A C_5H_{12} compound that gives four $C_5H_{11}Cl$ derivatives

   c.   A C_5H_{12} compound that gives only one $C_5H_{11}Cl$ derivative

   d.   Are there any other isomers of C_5H_{12}? Explain to each other how to find the answer to this question.

1

2.   Name each of the C_5H_{12} and $C_5H_{11}Cl$ molecules described in Problem 1.

3.   Build molecular models for each of the C_5H_{12} isomers.

4.   An experimental technique called ^{13}C nuclear magnetic resonance spectroscopy (NMR) allows chemists to tell how many different kinds of carbons are present in a molecule and whether carbons are primary (1°), secondary (2°), tertiary (3°), or quaternary (4°). Work together to find Kekulé structures for compounds having molecular formula C_6H_{12}. On each structure, identify carbons as 1°, 2°, 3°, or 4°; tell how many different kinds of carbons there are, and designate which carbons are equivalent.

    a.   A compound having only single bonds and only secondary carbons

    b.   A compound having only single bonds and primary, secondary, and tertiary carbons

    c.   A compound having only single bonds and primary, secondary, and quaternary carbons

    d.   A compound having only single bonds and primary, secondary, tertiary, and quaternary carbons

5. Divide into two groups. Work within your group to find as many structures as possible for constitutional isomers with molecular formula $C_3H_6O_2$. Draw Kekulé structures (show all bonds as lines and show all nonbonding electron pairs) for each isomer. Make sure that the following functional groups are represented: carboxylic acid, ester, ether, aldehyde, ketone, and alcohol. Circle each functional group and indicate its appropriate name. Finally, compare your structures with those discovered by the other group.

6. Propose the structure of a C_8H_{18} compound that gives only one $C_8H_{17}Cl$ derivative. Give the structure of the derivative as well. Explain your thought processes.

Reflection: Identify two incomplete or incorrect ideas about molecular structure, isomers, and functional groups that you brought to the Workshop. Explain how your understanding changed as a result of the Workshop. Consider whether you learned new thought patterns and problem-solving techniques. First develop your individual responses, then discuss your ideas as a group, and finally record the big "take-home points."

WORKSHOP 2

Structure: Molecular Geometry and Bonding

Purpose: The way atoms are connected is a fundamental aspect of molecular structure. The way atoms are arranged in space is also fundamental. Together, these two observables describe the structure of a molecule. There are different theories of chemical bonding, all of which must be consistent with the observed molecular structure. This Workshop starts with considerations of molecular geometry and then explores the use of valence bond theory to describe the observed molecular structure. As in Workshop 1, we can reason from observed properties to molecular structure. Also as in Workshop 1, we find that there are logical ways to solve problems.

Expectations: You should be prepared to discuss and use the following terms and concepts: Lewis structures, bonding and nonbonding electrons, formal charge molecular geometries, VSEPR, bond lengths, bond angles, hybridization, orbitals, and σ and π bonding.

1. The experimental observations for substituted methanes are as follows:
    (X,Y,Z = halogen atoms)

CH_2X_2	only one isomer identified
CH_2XY	only one isomer identified
CHXYZ	two isomers identified

    Consider two hypotheses about the molecular structure of substituted methanes:
    Square planar: The carbon is at the center of a square with the substituents at the corners.
    Tetrahedral: The carbon is at the center of a tetrahedron with the substituents at the corners.

    a. Work together to draw pictures of these two hypothetical structures. You will need to build a model of the tetrahedral structure in order to draw the picture.

5

1.   (continued)

   b. For the two hypothetical structures, use the drawings and the model to work out all
       the possible ways to arrange the substituents in space around the carbon:

       CH_2X_2

       CH_2XY

       $CH\,XYZ$

   c. Test the two hypothetical structures against the experimental observations. Explain
       your conclusions.

1.   (continued)

d. Would you bet a million dollars that one of the hypothetical structures is incorrect? Explain why or why not.

Would you bet a million dollars that one of the hypothetical structures is correct? Explain why or why not.

2.   Consider the experimental data about C-H bonds in the following table. Discuss the variations in bond dissociation energy, IR stretching frequency, and pK_a in terms of the observed geometries and our ideas relating structure and bonding. Be sure to explain the trends within the categories.

Molecule	∠ CCH	C-H BDE in kJ/mol	C-H BDE in kcal/mol	C-H IR stretch, cm^{-1}	C-H pK_a
Ethane	109.6°	420	100	2850-2960	~60
Ethylene	121.7°	444	106	3020-3100	44
Acetylene	180°	552	132	3300	25

Round robin is a technique in which each member of the group contributes sequentially to the solution of a problem. Start by assigning numbers to member of the group corresponding to the steps in the problem-solving format. For each new problem, change the assignment of individuals to steps.

"Structural" Round Robin

1. Start with the formula.
2. Decide how the atoms are connected (i.e., write a Kekulé structure). If there are several isomeric structures, choose one with which to proceed.
3. Write the complete Lewis structure, including nonbonding electron pairs.
4. Consider the geometry of the molecule, which is either (a) given or (b) guessed from VSEPR.
5. Choose the hybridization scheme that fits (a) the geometry and (b) the number of electron-pair units around the central atom.

$$\overset{\displaystyle a \quad\quad b}{\overset{\displaystyle \downarrow \quad\quad \downarrow}{\text{e.g., } CH_3C\equiv N:}}$$

    a. For a "central atom" of the methyl group, number of electron-pair units = 4.
    b. For a "central atom" of the nitrile group, number of electron-pair units = 2.

6. Overlap hybrid orbitals of the central atom with appropriate orbitals of the peripheral atoms, making the appropriate number of σ bonds.
7. Make π bonds as required to accommodate multiple bonds.
8. Place nonbonding pairs in orbitals.
9. Calculate formal charges. Check that octets are filled and that electrons are paired.
10. Label bonds as σ or π, identifying the orbitals that overlap to make the bonds (e.g., $\sigma(H1sCsp^3)$).

"Molecular Orbital" Round Robin

1.   n AO's $\rightarrow$ n MO's.

Construct MO's from the AO's in the absence of electrons.

Every bonding interaction (in phase) is accompanied by an antibonding interaction (out of phase). Those MO's or energy levels are always present, whether occupied or not.

2.   Place the available electrons with the MO's, two at a time (Pauli principle).

3.   Analyze bonding vs antibonding to estimate net bond energy

    (e.g., Compare H_2^+, H_2, H_2^-).

3. For each of the following molecules, give σ, π descriptions of the bonding that correspond to the observed structural information. Be sure to accommodate all of the bonding and nonbonding valence electrons and to specify any formal changes. Label the hybridization at the relevant atoms.

a. $H_aH_bC_aC_bH_cC_cC_dH_d$
vinylacetylene

$\angle H_aC_aC_b = 119°$ $H_aC_a = 111$ pm
$\angle H_bC_aC_b = 122°$ $H_dC_d = 109$pm
$\angle C_aC_bC_c = 123°$ $C_a\text{-}C_b = 134$ pm
$\angle C_bC_cC_d = 178°$ $C_b\text{-} C_c = 143$ pm
$\angle C_cC_dH = 182°$ $C_c\text{-}C_d = 122$ pm

H_a is *cis* to H_c

b. CH_3COCH_3
acetone

C-H = 109 pm
$\angle$ CCC = 116°
Atoms C,C,C and O are in the same plane

c. DHCCHBr
1-bromo-2-deuterioethylene

All atoms are in the same plane.
Two isomers are known.

3. (continued)

d. HCO_2H
   formic acid

$\angle$ HCO   = 124°   C-O = 120 pm
$\angle$ HC-OH = 111°   C-OH = 134 pm
$\angle$ OC-OH = 125°
All atoms are in the same plane.

e. $(H_3C)_2SO$
   dimethyl sulfoxide
   (DMSO)

$\angle$ CSO = 107°
$\angle$ CSC = 97°

The dihedral angle defined by the
CSC plane and the S-O bond = 116°.

f. CH_3NO_2
   nitromethane

$\angle$ CNO = 117°
$\angle$ CNO' = 117°   N-O = 122.4 pm
$\angle$ ONO' = 126°   N-O' = 122.4 pm

4. Construct molecular orbital diagrams and make qualitative predictions about the bond energies for He_2^+ and He_2. Explain why only one of the two species has been observed.

5. Three dicarboxylic acids ($C_2H_2(CO_2H)_2$) **I**, **J**, and **K** are catalytically hydrogenated (react with H_2 in the presence of a catalyst) to give dicarboxylic acids with formulas of $C_2H_4(CO_2H)_2$. Hydrogenation of both **I** and **J** gives the same dicarboxylic acid **L**. Compound **K** reacts with hydrogenates to form compound **M**. Give structures for compounds **I–M**. **Explain your reasoning.** First work as a group to find all of the experimental observations. Then discuss the deduction(s) you can make from each observation. Make multiple hypotheses about the structures as soon as you can. Be alert to ambiguity.

OBSERVATION   **DEDUCTION**

Reflection: Consider the difference between an observable and a theory. Discuss the difference with your teammates. Ultimately, summarize the difference in your own words.

In Problem 1, you evaluated multiple hypotheses. Discuss the value of starting with more than one hypothesis. For example, do you think you know more about the structure of tetravalent carbon compounds by having two hypotheses (square planar and tetrahedral) or just one (tetrahedral)? Explain how you used multiple hypotheses in solving Problem 5.

WORKSHOP 3

Structure and Properties

Purpose: The chemists' BIG IDEA is that the properties of a molecule are a consequence of the structure of the molecule. In this Workshop, we will explore the relationships of molecular structure to physical properties such as infrared absorption, dipole moment, boiling point, and solubility. Some of these are molecular properties—consequences of the structures of individual molecules. Others are intermolecular properties–consequences of the interactions between molecules. Of course, most of the intermolecular forces that determine the interactions between molecules have their origins in the structures of individual molecules.

We will work back and forth, deducing molecular structure from observable properties and, conversely, predicting properties from observable molecular structure.

Expectations: You should review and be prepared for an initial discussion about the meaning of the following key words and concepts: infrared absorption spectroscopy, dipole moments, intermolecular forces, van der Waals forces, dipole–dipole interactions, ion–dipole interactions, and hydrogen bonding.

1. The infrared spectrum of methanol (CH_3OH) vapor shows a sharp band around 3600 cm^{-1}, whereas the spectrum of liquid methanol shows a broad band at about 3400 cm^{-1}.

   a. Illustrate the differences between the arrangements of the methanol molecules in the liquid and gaseous states.

   b. Explain in words the difference between the arrangements and the properties of the methanol molecules in the liquid and gaseous states.

   c. Draw a picture showing your sense of the arrangements of the methanol and water molecules when 1 g of methanol is added to 1000 g water.

2.   For each of the given lettered compounds, give structures that are *consistent* with the
       following experimental observations:

       Two compounds, **G** and **H,** have the formula C_3H_6.   The infrared spectra for **G** and **H**
       are as follows:

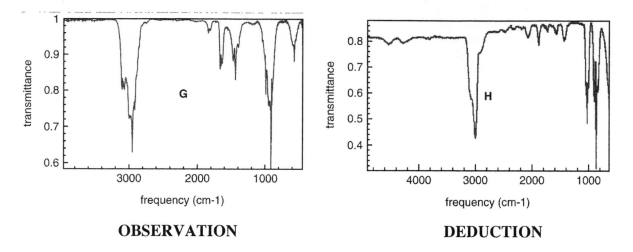

OBSERVATION   **DEDUCTION**

Spectra reprinted with permission from
"NIST Chemistry WebBook (http://webbook.nist.gov/chemistry)"

3.   For each of the compounds shown, designate the direction of the individual bond dipoles. (+➤) (Ignore C-C and C-H bonds.) Also, indicate the direction of the net molecular dipole moment. For each of the pairs of compounds shown, predict and explain which has the larger dipole moment. Then predict and explain which has the higher boiling point.

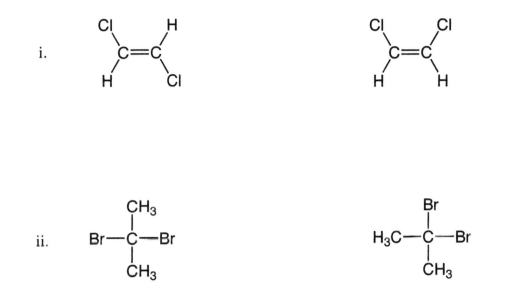

4.   Consider the two molecules CO_2 and SO_2. Experimentally, the net molecular dipole moment of one of them is zero, while the other one has a dipole moment of 1.61 D. Use your understanding of molecular structure to predict and explain which is which.

5.   Each of the following pairs of compounds are observed to have different boiling points. The experimental boiling points are given in the parentheses between each pair of compounds, but not necessarily in the correct order. In each case, circle which member of the pair has the higher boiling point and explain why. You might want to work this problem and the next one in round-robin format.

a. $(CH_3)_2CHCH(CH_3)_2$ (58°C, 69°C) $CH_3(CH_2)_4CH_3$

b. $(CH_3)_2CHOCH(CH_3)_2$ (35°C, 68°C) $CH_3CH_2OCH_2CH_3$

c. $CH_3\overset{\overset{\displaystyle O}{\|}}{C}NHCH_3$ (153°C, 204°C) $H\overset{\overset{\displaystyle O}{\|}}{C}N(CH_3)$

d. iodomethane   (174°C, 41°C)   decane
    (MW = 142)   (MW = 142)

16

6.   The boiling points of each member of the following pairs are specified in parentheses. For each member of each pair, make deductions about its molecular structure that are consistent with the experimental observations. In each case, explain your reasoning.

a. sodium chloride (1,413°C)   ethyl chloride (12°C)

b. $CH_3(CH_2)_3CH=CH_2$ (64°C) $CH_3(CH_2)_3CH=O$
(103°C)

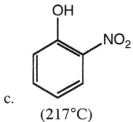

c.
(217°C)

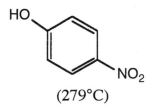

(279°C)

7. The members of each of the following pairs of compounds are observed to have different solubilities in water. For each pair, circle the more soluble compound and explain your choice.

    a. $CH_3CH_2OCH_3$ VS $CH_3CH_2SCH_3$

    b. $CH_3(CH_2)_3CH=CH_2$ VS $CH_3(CH_2)_3CH=O$

    c. $CH_3(CH_2)_4COOH$ VS $CH_3(CH_2)_4COONa$

    d.   iodomethane   VS   decane

8. a.   Qualitatively predict and explain the relative water solubility of the following compounds:

         i. pentane   vs   hexadecane

         ii. pentanoic acid vs   hexadecanoic acid

         iii. sodium pentanoate vs   sodium hexadecanoate

8.   (continued)

b. The sodium salts of long-chain fatty acids such as hexadecanoic acid are commonly called soaps. Although they mix almost completely with water, they have unusual solubility properties and do not dissolve as solvent-separated ions. Rather, they tend to aggregate in water to form discrete structures. Represented next are various ways in which the molecules of sodium hexadecanoate might interact, either at the surface of the water or in the bulk liquid phase. Analyze each case, considering all possible intermolecular interactions, to predict the relative stabilities of the four arrangements.

Let $CH_3(CH_2)_{14}CO_2^-$ Na^+ be represented by

Case 1

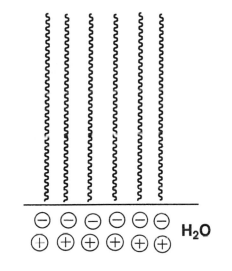

Case 2

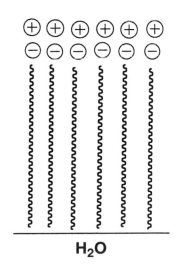

8.   (continued)

Case 3

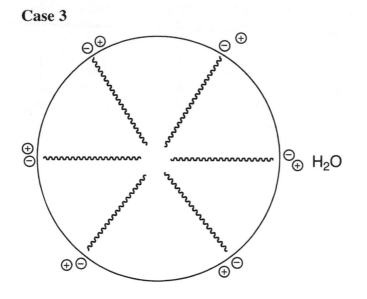

Case 4

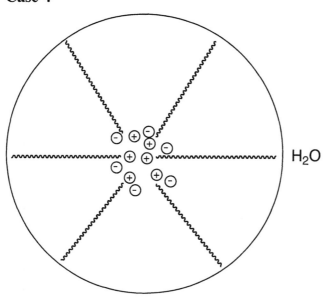

Reflection: Make a chart that categorizes each physical property as a characteristic of individual molecules or a consequence of interactions between molecules.

WORKSHOP 4

Structure and Properties: Acids and Bases

Purpose: "What reacts with what?" is a big question. One fundamental answer to this question is that acids react with bases. In this Workshop, you will explore a variety of acid–base reactions and the curved-arrow representation of the reactions that follows from Lewis acid–base theory. By the end of the Workshop, you will be able to predict the reactants or the products of acid–base reactions. Making those predictions requires that you understand the idea behind those reactions and not simply memorize $A + B \rightarrow C + D$.

Expectations: You should come to the Workshop with a good understanding of Lewis structures, nonbonding electron pairs, molecules with empty orbitals (unfilled octets), antibonding molecular orbitals, polar bonds, K_a, pK_a and Lewis acid–base theory.

1. Draw complete Kekulé–Lewis structures of reactants and products, using lines for bonds and showing nonbonding electron pairs. Identify the Lewis acid and the Lewis base for each reaction. Use curved arrows to show bond-making and bond-breaking processes for the forward and reverse reactions Explain the origins of the formal charges in the products.

$$CH_3SCH_3 + BF_3 \rightleftharpoons (CH_3)_2\overset{+}{S} - \overset{-}{B} F_3$$

$$(CH_3)_3\overset{+}{C} + H_2O \rightleftharpoons (CH_3)_3C - \overset{+}{O}H_2$$

$$CH_3SH + CH_3\overset{-}{O} \rightleftharpoons CH_3\overset{-}{S} + CH_3OH$$

$$\underset{CH_3\overset{\overset{O}{\|}}{C}CH_3}{} + H_3O^+ \rightleftharpoons H_2O + \underset{CH_3\overset{\overset{+OH}{\|}}{C}CH_3}{}$$

$$H_3N + CH_3 - Br \rightleftharpoons H_3\overset{+}{N} - CH_3 + \overset{-}{Br}$$

2.   The curved-arrow formalism tracks the movement of electrons and the bond-making and bond-breaking changes in an elementary step in a reaction mechanism. A more detailed view identifies the molecular orbitals that are involved. In general, the highest occupied molecular orbital (HOMO) of a Lewis base donates the electrons for the new bond, and the lowest unoccupied molecular orbital (LUMO) of a Lewis acid accepts the electrons. In most, but not all, cases, the process of accepting electrons leads to bond breaking. For each of the reactions that follow, use curved arrows to track the electrons and show the bond-making and bond-breaking processes. Identify the HOMO and LUMO involved in the processes. Finally, use the processes to predict the products of each reaction.

$H_3N + BF_3 \longrightarrow$

$H_3O^+ + NH_3 \longrightarrow$

$I^- + CH_3Br \longrightarrow$

$HCl + (CH_3)_2 C = CH_2 \longrightarrow$

$HO^- + (CH_3)_2C = O \longrightarrow$

3. For each of the following reactions, designate the acids and the bases, and use curved arrows to show the bond-making and bond-breaking processes as the reaction proceeds in the forward and reverse directions. Circle the major species at equilibrium. Start by drawing complete Kekulé–Lewis structures, showing all bonding and nonbonding electron pairs at the reaction centers.

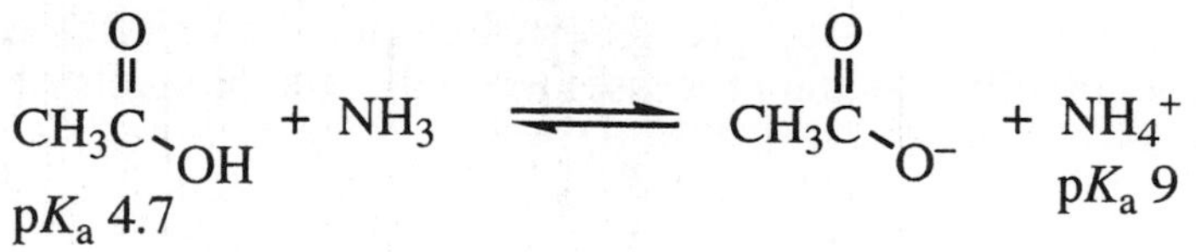

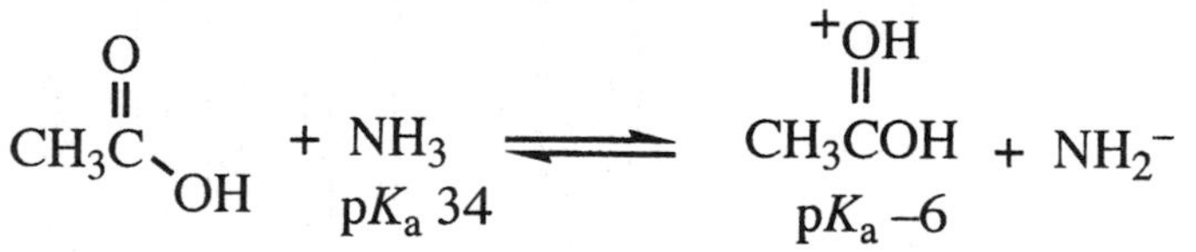

$$\underset{pK_a\ 4.7}{CH_3C\overset{O}{\underset{OH}{\parallel}}} + NH_3 \rightleftharpoons \underset{pK_a\ 9}{CH_3C\overset{O}{\underset{O^-}{\parallel}}} + NH_4^+$$

$$\underset{pK_a\ 34}{CH_3C\overset{O}{\underset{OH}{\parallel}}} + NH_3 \rightleftharpoons \underset{pK_a\ -6}{CH_3C\overset{^+OH}{\underset{}{\parallel}}OH} + NH_2^-$$

$$\underset{pK_a\ 15.5}{HCl + CH_3OH} \rightleftharpoons \underset{pK_a\ <-2}{H-\overset{+}{Cl}-H + CH_3O^-}$$

$$\underset{pK_a\ -2}{HCl + CH_3OH} \rightleftharpoons \underset{pK_a\ -2}{Cl^- + CH_3\overset{+}{OH}_2}$$

$$\underset{pK_a\ 24}{CH_3C\equiv CH + {}^-NH_2} \rightleftharpoons \underset{pK_a\ 34}{CH_3C\equiv C^- + NH_3}$$

$$\underset{pK_a\ 50}{CH_3CH_3 + {}^-NH_2} \rightleftharpoons \underset{pK_a\ 34}{CH_3CH_2^- + NH_3}$$

4.   Consider the relationship of pK_a to the free-energy change $\Delta G°$ in the reaction of an acid with water.

    a.   Start with a definition of pK_a.

    b. K_a describes a specific chemical reaction. Write a chemical equation for the reaction of the ammonium ion that occurs when $NH_4^+Cl^-$ is dissolved in water.

    c.   Give a mathematical expression for K_a.

    d.   Explain how K_a is related to $\Delta G°$, the change in free energy that occurs for the reaction of the ammonium ion when $NH_4^+Cl^-$ is dissolved in water.

    e.   Explain qualitatively how ΔpK_a (the pK_a of the starting acid minus the pK_a of the product acid) describes the following more general equilibrium:

$$HA + B \rightleftharpoons BH^+ + A^-$$

    f.   Explain qualitatively the origin of the following equilibrium mnemonics:
"The stronger base wins the proton."
"The weaker acid predominates."
"Proton affinity tracks with pK_a. "

5. a.   Predict the product(s) for the following reactions. The ideas that you used in the previous problems will be useful.

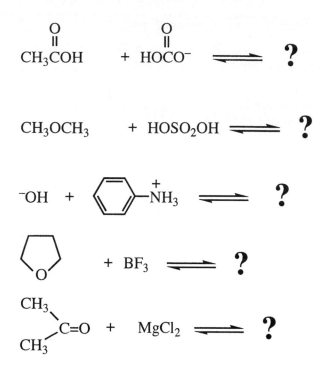

$$CH_3\overset{\overset{O}{\|}}{C}OH \quad + \quad HO\overset{\overset{O}{\|}}{C}O^- \quad \rightleftharpoons \quad ?$$

$$CH_3OCH_3 \quad + \quad HOSO_2OH \quad \rightleftharpoons \quad ?$$

$$^-OH \quad + \quad \langle \rangle -\overset{+}{N}H_3 \quad \rightleftharpoons \quad ?$$

$$\underset{O}{\langle \rangle} \quad + \quad BF_3 \quad \rightleftharpoons \quad ?$$

$$\underset{CH_3}{\overset{CH_3}{>}}C=O \quad + \quad MgCl_2 \quad \rightleftharpoons \quad ?$$

b.   Choose the reactants to form the following products:

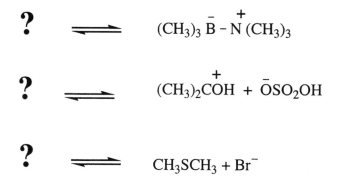

$$? \quad \rightleftharpoons \quad (CH_3)_3 \overset{-}{B} - \overset{+}{N}(CH_3)_3$$

$$? \quad \rightleftharpoons \quad (CH_3)_2\overset{+}{C}OH \quad + \quad {}^-OSO_2OH$$

$$? \quad \rightleftharpoons \quad CH_3SCH_3 + Br^-$$

6. Two isomeric compounds, **A** and **B**, are known to each have a monosubstituted benzene ring (C_6H_5-). Both have the formula $C_6H_5C_3H_5O_2$ and both are insoluble in water. However, **A** dissolves in dilute aqueous NaOH, but **B** does not. **A** has an intense, broad IR absorption at 3000–2500 cm^{-1}; this absorption is absent in the IR spectrum of **B**. Both **A** and **B** have strong IR absorptions in the region 1750 –1650 cm^{-1}. Give structures for **A** and **B** *consistent* with this information. **Explain your reasoning.** Is the information provided *sufficient* to uniquely define the structures **A** and **B**?

OBSERVATION   DEDUCTION

Useful Acidity Constants

$$K_a = \frac{[H_3O^+][A^-]}{[HA]} \qquad pK_a = -\log_{10}K_a$$

	Acids	pK_a	Conjugate Bases
Strong $pK_a < 0$	$HOSO_2OH$	−5.2	$HOSO_2O^-$
	HI	−5.2	$^-\!:\!I$
	HBr	−4.7	$^-\!:\!Br$
	HCl	−2.2	$^-\!:\!Cl$
	H^+OH_2	−1.7	$:OH_2$
	$HONO_2$	−1.4	$^-\!:\!ONO_2$
$pK_a \sim 5$	$HOSO_2O^-$	1.99	SO_4^{-2}
	HF	3.18	$^-\!:\!F$
	$HOCCH_3$ (C=O)	4.75	$^-\!:\!OCCH_3$ (C=O)
	$HOCOH$ (CO_2 in H_2O) (C=O)	6.4	$^-\!:\!OCOH$ (C=O)
$pK_a \sim 10$	H^+NH_3	9.25	$:NH_3$
	HO—〈benzene ring〉	10.0	$^-\!:\!O$—〈benzene ring〉
	$HSCH_2CH_3$	10.6	$^-\!:\!SCH_3$
$pK_a \sim 15$	$HOCH_3$	15.5	$^-\!:\!OCH_3$
	HOH	15.7	$^-\!:\!OH$
	$HOCH_2CH_3$	15.9	$^-\!:\!OCH_2CH_3$
$pK_a \sim 25$	HCH_2CCH_3 (C=O)	20	$^-\!:\!CH_2CCH_3$ (C=O)
	$HC\equiv CH$	24	$^-\!:\!C\equiv CH$
	HNH_2	34	$^-\!:\!NH_2$
$pK_a \sim 50$	$HCH=CH_2$	45	$^-\!:\!CH=CH_2$
	HCH_3	~60	$^-\!:\!CH_3$

Reflection:

a. Explain to your Workshop colleagues how the following types of reactions are subcategories of the prototypical Lewis acid–base reaction:

Lewis Prototype:

$$H_3N: + BF_3 \rightarrow H_3\overset{+}{N}\text{-}\overset{-}{B}F_3$$

Brønsted Acid–Base Reaction:

$$H_3N: + HA \rightarrow H_3\overset{+}{N}\text{-}H + A\overset{-}{:}$$

Organic Reaction:

$$H_3N: + CH_3\text{–}Br \rightarrow H_3\overset{+}{N}\text{–}CH_3 + Br^-$$

b. Write a short coherent paragraph that explains how the curved-arrow formalism is a graphical representation of the Lewis acid–base theory of chemical reactions.

WORKSHOP 5

Reaction Mechanism

Purpose: Expert students have organized their knowledge. The structure of their understanding helps them learn and retain new knowledge and helps them access what they have learned. It is one level of knowledge to learn the reactants and products for characteristic chemical reactions. This empirical knowledge might even be organized according to the nature of the change in the reaction: conformational change, an addition reaction, a substitution reaction, etc. In the end, however, it remains empirical classification.

It is an altogether different level of organization and understanding to be able to think about *how* reactants are transformed to products. A description of the bond-making and bond-breaking path from reactants to products is called a reaction mechanism. Modern organic chemistry is marvelously rational, coherent, and organized because of our understanding of the mechanism of organic reactions. This Workshop should help you organize your developing knowledge about chemical reactions. The application of fundamental ideas about acids, bases, polar bonds, and chemical reactions will lead you to rational *predictions* about the mechanisms of reactions that you have not seen before. In a subsequent Workshop (Workshop 15), we will test those predictions against experimental observations.

Expectations: You should review and understand the *subsidiary concepts* in Problem 1.

1. Work together with your colleagues to construct a concept map that organizes and summarizes our ideas about *chemical reactions*. Use linking verbs and phrases to show how subsidiary concepts are connected to the central concepts of *chemical reactions*.

    For example,

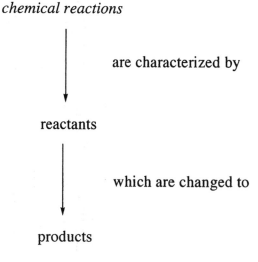

1.   (continued)

Following is a list of subsidiary concepts. Start by making a rough ranking of the importance of these concepts. That will help you construct a hierarchical map. One useful way to work on the map is to apply self-adhesive notes so that the map can be easily changed as it evolves.

Subsidiary Concepts

Products	Structure	Reactivity
Change	Acid–Base	LUMO/HOMO
Electron Deficient	Bonds	Polar Bonds
Formal Charge	Curved Arrows	Nonbonding Electrons
Intermediates	Pi Bonds	δ^*, π^*
Electron Rich	Bond Breaking	Bond Making
Sigma Bonds	Electrophile	Nucleophile
Transition State	Incomplete Octets	Reactants

2. **Writing Reaction Mechanisms.** A reaction mechanism is a detailed hypothesis about the way a reaction occurs. A mechanism is a *deduction* based on experimental *observation* about the reaction. Some reaction mechanisms are well established. In other cases, the mechanism may be speculative and is likely to change as more data become available. Mechanisms map the path by which the reactants change into products and the movement of electrons that accompanies this change. They show how reactants interact, the final transformation to products, the structures of transition states, and the intermediates that are formed. Each step in a multistep mechanism is an elementary step with a single transition state. An elementary step can involve bond making, bond breaking, or both. Bond breaking and making are shown with the help of curved arrows.

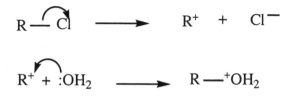

In this problem, we will consider the reactions of a three-membered cyclic ether, also known as an epoxide, with water to give a 1,2-diol. This ring-opening reaction takes place in an aqueous solution under acidic or basic conditions. The ring opening is very slow in water at pH 7. We will work using our ideas about chemical reactions to construct a hypotheses about mechanisms for these reactions. The round-robin format will work well for this problem.

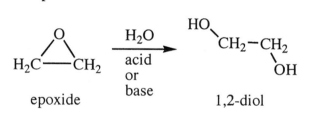

2.   (continued)

a.   First consider basic conditions:

(1)   Which reactive species are present in a basic solution in water?

(2)   What is the charge on the reactive species?   Characterize the reactive species as Lewis acids or Lewis bases.

(3)   Identify the possible site(s) of reactivity in the epoxide molecule. Characterize these sites as Lewis acids or Lewis bases.

(4)   With the help of arrows, show the bond-making and bond-breaking processes that correspond to a reaction between the epoxide molecule and the reactive species.

(5)   Write the structure of the first product formed in the reaction proposed in (4).

(6)   Is this intermediate product cyclic or an open chain?

(7)   What will be the charge on this intermediate? Is it a Lewis acid or a Lewis base?

(8)   Show, with the help of arrows, how the free diol is formed from this intermediate.

(9)   Is this reaction catalytic? Explain.

2.   (continued)

      b.   Now consider acidic conditions:

          (1)   Which reactive species are present in an acidic solution in water?

          (2)   What is the charge on the reactive species? Characterize the reactive species as Lewis acids or Lewis bases.

          (3)   Identify the possible site(s) of reactivity in the epoxide molecule. Characterize these sites as Lewis acids or Lewis bases.

          (4)   With the help of arrows, show the bond-making and bond-breaking processes that correspond to a reaction between the epoxide molecule and the reactive species.

          (5)   Write the structure of the first product formed in the reaction proposed in (4).

          (6)   Is this intermediate product cyclic or an open chain?

          (7)   What will be the charge on this intermediate? Is it a Lewis acid or a Lewis base? Identify the possible site(s) of reactivity in this intermediate.

          (8)   Identify a species in solution that would react with this intermediate.

          (9)   Show, with the help of arrows, how the intermediate reacts with a species in solution. (Remember that the net reaction involves ring opening.)

2.   (continued)

(10) Write the structure of the first product formed in the reaction proposed in (9).

(11) What is the charge on this first-formed product?

(12) Show, with the help of arrows, how the free diol is formed from this first-formed product.

(13) Is this reaction catalytic? Explain.

c.   Make a chart that summarizes the acids (electrophiles) and the bases (nucleophiles) in the ring-opening steps for the base- and acid-catalyzed reactions.

Reflection: Summarize the key concepts about chemical reactions that you used to construct predictions about the mechanisms of reaction of an epoxide with water in an acid and a basic solution.

WORKSHOP 6

Stereochemistry of Alkanes and Cycloalkanes

Purpose: We return once again to issues of isomerism and the geometric structure of molecules. Pairs of isomers are related as either constitutional isomers or stereoisomers. In this Workshop, we explore the fascinating world of stereoisomers: isomers that differ only in the way their atoms are arranged in space. You will learn to identify different stereoisomers, to recognize configurational and conformational stereoisomers, and to evaluate energy differences between different stereoisomers. Finally, we will analyze the change of one conformational isomer into another as a simple chemical reaction, with fundamental characteristics in common with more complex reactions.

Expectations: You should be prepared to discuss and use the following terms and ideas: IUPAC and CIP nomenclature; isomers; constitutional isomers; stereoisomers; *cis-* and *trans-*; configuration; conformation; *anti, gauche,* and *eclipsed*; sawhorse and Newman representations; chair conformation; equatorial; axial; equilibrium; rate; $\Delta G°$; and $\Delta G^{\neq}$.

1. Work with a partner to specify the rules of IUPAC and CIP nomenclature for alkenes. Apply those rules by giving the proper name for the following compound, including specification of the double-bond stereochemistry as Z or E:

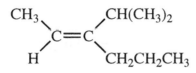

2. Draw all isomeric dimethylcyclobutanes, clearly showing their stereochemistry. Name each isomer. Indicate whether pairs of structures are related as constitutional isomers or stereoisomers, and explain how the names reveal the relationships.

WORKSHOP 6 STEREOCHEMISTRY OF ALKANES AND CYCLOALKANES

3. a. For *n*-butane, make molecular models of the *anti* and *gauche* conformations, as well as the conformation in which the two methyls are eclipsed. Using your molecular model, measure the distance between the methyl carbons in the *anti, gauche,* and *eclipsed* conformational isomers. Make careful sawhorse drawings of the three conformational isomers.

    b. Make a molecular model of methylcyclohexane, and carefully draw the two chair conformations. Measure the distance between the methyl carbon and the carbons at the 3- and 5- positions of the ring. Compare these measurements with those you previously made on butane. On the basis of your measurements, which of the two cyclohexane conformations do you predict will predominate at equilibrium? Is the methyl group *axial* or *equatorial* in the more stable conformation? **Explain your reasoning.**

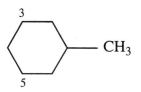

    c. The *gauche* conformation of *n*-butane is 0.9 kcal/mol higher in energy than the anti isomer. Based on your comparison of butane and methycyclohexane, estimate the energy difference between *axial-* and *equatorial*-methylcyclohexane. Explain your reasoning.

38

4. a. Using only chair conformations, make and draw molecular models for all of the 1,4–dibromocyclohexanes. Indicate clearly which molecules are related as conformational isomers and which are related as configurational isomers. Designate the configurational isomers as *cis* or *trans*. **Explain** which isomers equilibrate rapidly with each other (interconvert) at room temperature.

    b. Build a molecular model of the chair conformation of *trans*-1,4-dibromocyclohexane. Then construct a qualitative reaction-coordinate diagram that describes the energy of the system as a function of the progress of the interconversion of one chair conformational isomer into the other. (Ignore any nonchair energy minima along the pathway from one chair conformation to the other.) On the diagram, be sure to clearly indicate the coordinates that correspond to the different chair conformations. Specify the energy difference that controls the population of the two conformations at equilibrium. Using a model of 1,4-dibromocyclohexane, **explain (in words)** how your diagram corresponds to changes in the molecular structure of the compound.

5.   Consider rotations around the C2-C3 bond in 2-methylbutane.

    a.   Represent all the staggered conformations in Newman projections.

    b.   Specify the relative energies of the staggered conformations.

    c.   Represent the eclipsed conformations that must be traversed to interconvert the staggered conformations.

    d.   Specify the relative energies of the eclipsed conformations.

    e.   Construct a qualitative plot (reaction-energy diagram) that describes the energy of the system as a function of the progress of the interconversion of the staggered conformational isomers.

    f.   On the diagram, be sure to clearly indicate the coordinates that correspond to the different conformations.

    g.   Specify the energy differences that control the populations of the staggered conformations at equilibrium.

    h.   Specify the energy differences that control the rate of the interconversion reactions. (There are forward and backward reaction rates.)

    i.   Using a model, explain (in words) how your reaction-energy diagram describes changes in energy as a function of changes in molecular structure.

Reflection: Write a short paragraph that explains clearly how the energy difference between the *gauche*, *anti*, and *eclipsed* conformations of butane provides the basis for analyzing energy differences between the conformational isomers of more complex structures such as methylcylohexane and 2-methylbutane.

WORKSHOP 7

Alkenes: Electrophilic Addition Mechanism: Carbocations

Purpose: At first glance, there seems to be a bewildering array of chemical reactions of the carbon–carbon double bond. That is why those compounds constitute a useful and interesting functional group. On the other hand, you probably wonder how to make sense of (learn) these reactions. In this Workshop, our purpose is to recognize the recurring patterns and to understand the regiochemical results and common mechanism of a wide variety of addition reactions to carbon–carbon double bonds. Simultaneously, you should come to understand how and why chemists choose one mechanistic hypothesis over another.

Expectations: Prior to this Workshop, review acid–base chemistry, Lewis structures, reaction energetics ($\Delta G°$, $\Delta G^{\neq}$), electrophiles, nucleophiles, addition reactions, reaction mechanisms, reaction-energy diagrams, and catalysis.

1. For each of the following pairs of isomers, circle the *major* species at equilibrium. **Explain your choices convincingly**. (Since this question is about thermodynamics, you do not have to worry about how the reaction occurs or whether it occurs at a significant rate.)

    a. (Z)-2-butene ⇌ (E)-2-butene

    b.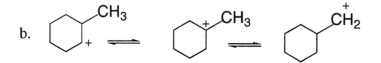

    c. Qualitatively specify the energy changes $\Delta G°$ for each of the transformations in parts a and b.

42

2. For each of the following pairs of reactions, circle the reaction that is faster. Be careful to explain your choices.

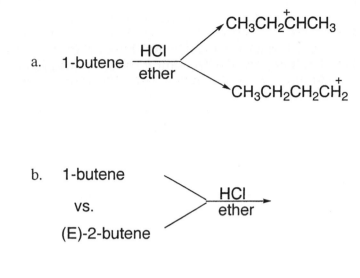

a. 1-butene $\dfrac{\text{HCl}}{\text{ether}}$ $\nearrow CH_3CH_2\overset{+}{C}HCH_3$
$\searrow CH_3CH_2CH_2\overset{+}{C}H_2$

b. 1-butene

    vs.

    (E)-2-butene

    $\dfrac{\text{HCl}}{\text{ether}}$

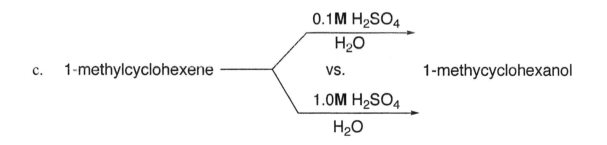

c. 1-methylcyclohexene

    $\dfrac{\text{0.1M H}_2\text{SO}_4}{\text{H}_2\text{O}}$

    vs.   1-methycyclohexanol

    $\dfrac{\text{1.0M H}_2\text{SO}_4}{\text{H}_2\text{O}}$

d. Qualitatively specify the relative activation energy $\Delta G^{\neq}$ for the competing reactions in parts a, b, and c.

3.   Consider the following set of reactions. Reagents are shown above, solvents below, the
reaction arrows.

a.   Work on this first part of the problem using the entire team in round-robin
format. Make a table in your book listing the electrophile and the nucleophile
that added to the double bond in each of the reactions in the set.

b.   Now switch to a pairwise format for Parts b and c. Consider the first three
reactions in the set. Explain how the formation of different products under
different reaction conditions reveals a reactive intermediate on the mechanistic
pathway.

c.   Write a single generalized mechanism in chemical equations that explains the
entire set of reactions. Ultimately, share your generalized mechanism with the
other pairs in your team.

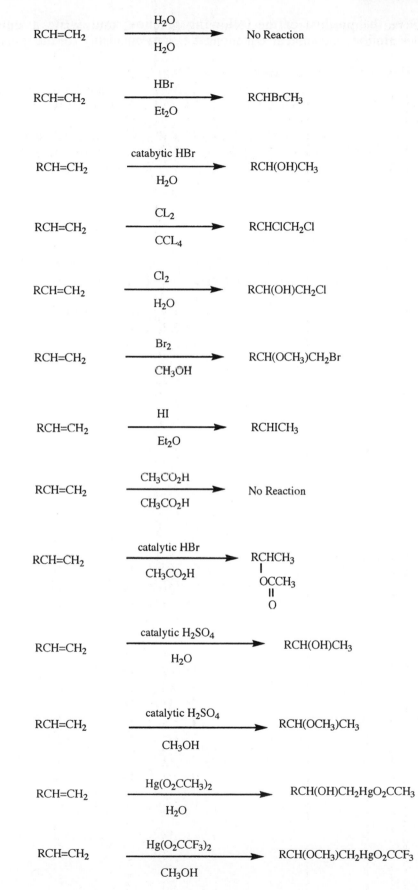

4.   a.   Give the product of the following reaction, and write a detailed stepwise description, in chemical equations (i.e., a mechanism), for the transformation.

$(CH_3)_2C=C(CH_3)_2 + HCl \rightarrow$

b.   Identify each species in the mechanism as a Lewis acid or a Lewis base, and use curved arrows to describe bond-making and bond-breaking processes.

c.   Construct a plot that describes the energy of the system as a function of the progress of the reaction (a reaction-energy diagram).

d.   On your diagram, specify the energy difference that corresponds to the $\Delta G°$ for the overall reaction and the energy difference that corresponds the $\Delta G^{\neq}$ for each step.

e.   **Explain (in words)** what is happening as the system makes its way from reactants to products.

5.   The addition of HBr to 2-methylpropene can, in principle, give either 1-bromo-2-methylpropane or 2-bromo-2-methylpropane. In practice, only the latter is formed.

  a.   Write mechanisms for the formation of each of these alkyl bromides, using curved arrows to indicate the bond-making and bond-breaking processes.

  b.   Now construct two reaction-energy diagrams that describe mechanisms for the formation of each alkyl bromide.

  c.   Compare these two reaction-coordinate diagrams, and **explain (in words)** why 2-bromo-2-methylpropane is the only product formed.

Reflection: In sum, discuss and explain why the idea of a reaction mechanism is a BIG IDEA. This Workshop is also about an important THINKING SKILL. Discuss and explain how one chooses between competing hypotheses.

WORKSHOP 8

Alkenes: Reactions

Purpose: To understand the range of addition reactions to carbon–carbon double bonds and to explore the synthetic utility of these reactions.

Expectations: Prior to this Workshop, review the reactions and mechanisms of additions to alkenes. Also review stererochemistry (enantiomers and diastereomers), conformational analysis, steric effects (Van der Waals repulsions), carbocations, equilibrium, and rates of reaction.

1. Give the products(s) for each of the following reactions, clearly showing stereochemistry. In each case, clearly explain the mechanistic origin of the stereochemistry that you have specified.

    a.   *trans*-3-hexene

$$\xrightarrow{\text{cold dilute KMnO}_4}$$

or

1. OsO_4
2. $NaHSO_3$, H_2O

    b.

$$\xrightarrow[\text{2. } H_2O_2,\ NaOH/H_2O]{\text{1. } BH_3,\ Et_2O}$$

2. When one of the stereoisomeric ethylene dicarboxylic acids is reacted with Cl_2 in H_2O, the corresponding chlorohydrin is formed with the following stereochemistry:

$$HO_2CCH=CHCO_2H \xrightarrow[H_2O]{Cl_2}$$

Show a detailed mechanism for this reaction, using curved arrows to indicate the movement of electron pairs. Deduce the correct stereochemistry for the starting ethylene dicarboxylic acid. **Explain carefully** how your mechanism accounts for the stereochemistry of the reactant and product.

3. Give a reasonable mechanism (use curved arrows to show bond making and bond breaking) for the acid-catalyzed polymerization of 2-methylpropene. You need to show just a few of the polymer chain-lengthening steps.

$$(CH_3)_2C=CH_2 \xrightarrow[\substack{trace \\ H_2O}]{BF_3}$$

n = thousands

4. Propose structures for each of the lettered compounds that follow: Two isomeric compounds, **A** and **B**, have the molecular formula C_6H_{12}. Both **A** and **B** react with H_2 in the presence of PtO_2 catalyst to give 2,3-dimethylbutane. When **A** is heated with a few drops of concentrated H_2SO_4, it is converted to a mixture of **A** and **B** in which **B** predominates.

OBSERVATION   **DEDUCTION**

5. Compounds **C, D,** and **E** have the same molecular formula: C_4H_8. They all react with H_2/PtO_2 to give the same compound. The reaction of **C** or **D** with H_2O/H_2SO_4 or with BH_3-THF, followed by treatment with a basic solution of hydrogen peroxide, gives the same compound, namely, **F**. The reaction of **E** with H_2O/H_2SO_4 also gives compound **F**. However, the reaction of **E** with (1) BH_3-THF and (2) ^-OH, H_2O_2 gives a new compound, **G**.

OBSERVATION   **DEDUCTION**

6.   **H** and **I** are isomers with molecular formula C_4H_8. **H** reacts with H_2 in the presence of a catalyst to give an alkane, C_4H_{10}. **I** also reacts with H_2 under the same conditions to give C_4H_{10}, which is different from the compound obtained for **H**. The reaction of **H** with O_3, followed by treatment with Zn/H_3O^+, gives two products: CH_2O and C_3H_6O. **I** reacts under the same ozonolysis conditions to give one product: C_2H_4O. When **I** is heated with a few drops of H_2SO_4, it is converted to a mixture of **I, J,** and **K** in which **J** predominates. **I, J,** and **K** are all reduced with H_2/PtO_2 to the same alkane. Ozonolysis of **J** gives one product, C_2H_4O, the same product obtained from ozonolysis of **I**. Ozonolysis of **K** gives two products: CH_2O and C_3H_6O. This C_3H_6O compound from **K** is not the same as the C_3H_6O compound from **H**.

OBSERVATION   **DEDUCTION**

7.   We learn chemical reactions and spectroscopic techniques so that we can put them to use in solving problems.

Chemists routinely acquire information about molecular structure so that they know with which molecule they are working. For example, a chemist runs a reaction and needs to find out whether the material he or she isolated at the end of the reaction is starting material or product.

For the pairs of compounds that follow, propose simple (chemical or spectroscopic) laboratory tests that would tell you which of the two compounds you had. Explain exactly what you would do and exactly what you expect to observe.

a.   a cycloalkane vs. an alkene
(give a chemical method and a spectroscopic method)

b.   1-alkene vs. 1-alkyne

c.   cyclohexanol vs. hexanoic acid

8. Chemists routinely change one functional group into another, a key activity in making new molecules. To do this, the chemist needs to be able to think about how to effect a specific change.

    Propose chemical reactions that accomplish the objectives that follow. Specify reagents, essential conditions, and catalysts.

    a. Convert a *cis*-alkene to a mixture in which the more stable *trans*-alkene predominates.

    b. Switch the regiochemistry of an alkyl halide (e.g., convert 1-chloropentane into 2-chloropentane).

8.   (continued)

c.   Convert 3-ethyl-1-pentene into each of the three alcohols listed next. Be sure to control the regiochemistry.

$(CH_3CH_2)_2CHCH_2CH_2OH$

$(CH_3CH_2)_2CHCH(OH)CH_3$

$(CH_3CH_2)_3COH$

Reflection: Explain how an understanding of the mechanism of a reaction helps you learn the regiochemical and sterochemical characteristics of that reaction. Illustrate your explanation with examples from this Workshop.

Identify two ideas that were clarified by this Workshop. Be explicit about your understanding before and after (i.e., be specific about the changes in your thinking). Explain these changes to your teammates.

WORKSHOP 9

Free-Radical Reactions: Thermochemistry

Purpose: A second big answer to the big question "What reacts with what?" is that radicals react with radicals and with σ and π bonds. In this Workshop, you will explore reactions of molecules with unpaired electrons, distinguishing those reactions from the polar reactions of electron pairs. Characteristic properties of free radicals will be explored and distinguished from those of ions. You should be able to write chain reaction mechanisms and use bond dissociation energies to calculate ΔH for overall reactions and for individual mechanistic steps, allowing you to estimate relative rates of competing reactions.

Expectations: You should review and be prepared for an initial discussion about the meaning of the following key words and concepts: bond dissociation energies, ΔH, E_a, and ΔH_f; homolysis and heterolysis; reactivity, stability, and selectivity; and chain reactions—initiation, propagation, and termination.

1.   Hydrogen chloride can be dissociated in a number of different ways:

$$HCl \xrightarrow[\text{phase}]{\text{gas}} H\bullet \ + \ Cl\bullet \qquad \Delta H = +103 \text{ kcal/mol}$$

$$HCl \xrightarrow[\text{phase}]{\text{gas}} H^+ \ + \ Cl^- \qquad \Delta H = +333 \text{ kcal/mol}$$

$$HCl \xrightarrow[\text{solutions}]{\text{aqueous}} H^+_{aq} \ + \ Cl^-_{aq} \qquad \Delta H = -10 \text{ kcal/mol}$$

a.   Explain why the ΔH values are so drastically different in these three examples.

b.   Predict an approximate ΔH value for the following reaction in the gas phase:

$$HCl \ + \ H_2O \longrightarrow H_3O^+ \ + \ Cl^-$$

Explain your reasoning.

2.   The reaction of methanol with HBr follows a polar ionic, not a radical, mechanism:

$$CH_3OH + HBr \longrightarrow CH_3Br + H_2O$$

a.   Is it appropriate to use homolytic bond dissociation energies to calculate ΔH for this reaction? Explain.

b.   Look up the necessary homolytic bond dissociation energies in your textbook, and determine ΔH for this reaction. Give an enthalpy reaction diagram that corresponds to your calculations. Plot enthalpy on the vertical axis and starting material and product on the horizontal axis.

c.   Use the data that follow on heats of formation to calculate ΔH for the reaction. Give an enthalpy diagram that corresponds to your calculations.

ΔH_f values in kcal/mol: methanol (-48.1), methyl bromide (-8.5), HBr (-8.7), water (-57.8).

3. Ethylene can be polymerized at high temperatures and pressures by a free-radical chain reaction using a typical radical initiator, di-*tert*-butyl peroxide.

    a. Write an initiation step, at least three propagation steps, and a possible termination step for this reaction.

    b. Consider the bond changes, and evaluate whether the propagation step is exothermic or endothermic.

    c. Small amounts of branching are often found. In particular, the polymer chain is found to have occasional ethyl or butyl groups as side chains. Consider possible diversions of the normal propagation steps that might lead to these structural features.

    d. The polymerization of ethylene using Ziegler–Natta or other organometallic coordination catalysts leads to a polymer with very little branching. Such polymers are high-density polyethylene (HDPE, recycle code 2, e.g., milk jugs), while branched polymers are low-density polyethylene (LDPE, recycle code 4, e.g., sandwich bags). Discuss why these differences in structure lead to differences in physical properties and applications.

4.   The photochlorination of isobutane gives two monochlorination products: isobutyl chloride and *tert*-butyl chloride.

     a.   Write balanced equations for the formation of both products.

     b.   Give a chain mechanism for the reaction. Indicate clearly the competing reactions that lead to the two products.

     c.   Calculate ΔH for each step in your mechanism.

     d.   The actual product yield is 68% isobutyl chloride and 32% *tert*-butyl chloride. Give a potential-energy–reaction-coordinate diagram for the rate-determining steps that lead to the two products. Compare the two pathways in a single diagram, making sure to show the enthalpy changes and the relative energies of the two transition states.

4.   (continued)

    e.   Calculate ΔH for each step in the analogous photobromination chain
            mechanism.

    f.   The actual product yield is >99% *tert*-butyl bromide and a trace of isobutyl
            bromide. Give a potential-energy-reaction-coordinate diagram for the rate
            determining steps that lead to the two products. Compare the two pathways
            in a single diagram, making sure to show the enthalpy changes and the relative
            energies of the two transition states.

    g.   Compare your diagrams for chlorination and bromination to explain the
            observed difference in selectivity.

5.   Free-radical reactions are used in synthetic reactions, such as the two cited next.
      Propose reaction mechanisms, with an emphasis on the chain-propagating steps.

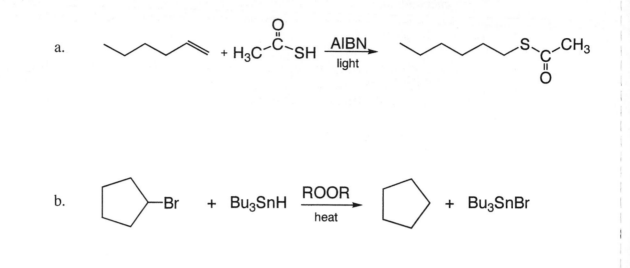

a.

b.

6.   The photochemical iodination of alkanes by iodine is not synthetically useful for
      virtually all alkanes. Consider whether tert-butyl hypoiodite (^{t}BuOI) would be a
      reasonable reagent for iodination of a typical alkane. Write a free-radical chain
      mechanism and calculate ΔH for all steps. For initiation, use dissociation of the O-I
      bond (48 kcal/mol). The O-H bond of tert-butyl alcohol has a bond dissociation
      energy of 105 kcal/mol.

Reflection: What is the reference state when ΔH is calculated from bond dissociation energies? Explain why ΔH (BDE) is always positive, but ΔH_f can be positive or negative. Identify at least two ideas about ΔH which you brought into this workshop that were incorrect or incomplete. Explain how your ideas changed as a result of the workshop. First develop your individual responses, then discuss them as a group, and then write out what you gained from the workshop.

When are radical reactions particularly useful for synthetic transformations? What limitations are there? (*Hint*: Think of radical reactions that do not work. Why do they fail?) Could the two transformations in Problem 5 have been accomplished by other, more traditional reactions?

WORKSHOP 10

Organic Synthesis

Purpose: Organic chemists make molecules. They make molecules that occur in nature. They also invent and make molecules that are the products of the chemists' imagination and curiosity (i.e., molecules that do not occur in nature). The first step in synthesizing (making) a more complicated molecule from simpler molecules is to design a plan. This design stage requires cognitive skills that are at the highest levels of Bloom's hierarchy of mental skills: application, analysis, synthesis, and evaluation. One of the primary reasons to learn the chemical reactions of the functional groups is because the reactions are the tools of organic synthesis. We can group the reactions into two general types: functional group interchanges and bond-making (or -breaking) reactions. We will be especially interested in ways to make carbon–carbon bonds, because doing that is central to making larger molecules from smaller building blocks.

Synthesis is fun because it is a creative act. This Workshop will give you opportunities to apply the chemical reactions you have learned to the synthesis of organic molecules. The problems are challenging because the goal is to get the desired constitutional isomer and the desired stereoisomer. Ultimately, we will explore retrosynthetic analysis, a step-by-step procedure for constructing a rational design of a synthetic plan. This takes practice because we tend to learn reactions in the forward direction (A→B→C).

Expectations: You need to have **mastered** starting materials, reagents, conditions, and products for the fundamental organic reactions of alkenes, alkyl halides, and alkynes in order to be able to use these reactions in synthesis. It will help to organize the reactions into functional group interchanges and bond-making reactions and to group together reactions that give products with well-defined stereochemistry. You should review the meaning of *cis*, *trans*, (E), (Z), chiral, achiral, and racemic mixture.

1. Propose methods for accomplishing the functional group interchanges in Chart 1. More than one step may be required. Specify reagents and reaction conditions for each step.

    Working together as a brainstorming team, simplify Chart 1 by identifying compounds that can serve as branch points to make several target molecules. Do this by thinking "backwards" to identify the branch points. For example, work your way around the perimeter of the molecule, considering the target molecules one by one. For each target, identify a precursor molecule that you could convert to the target in one step, using a reaction that you know. After you have done this for several target molecules, you will begin to see that there are common precursors (branch points) from which you can prepare several target molecules. Now the problem simplifies to preparing the precursors from chlorobutane. (The ⇐ symbol identifies a retrosynthethic step; it means that the target can be made from the precursor by a known reaction.)

64

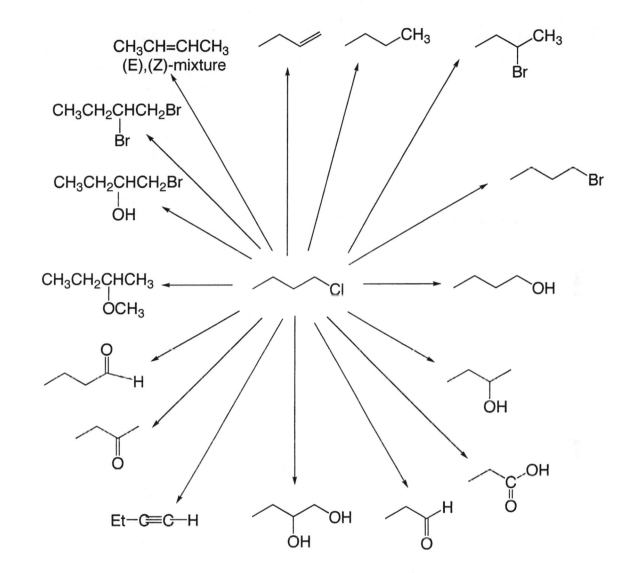

Chart 1

2. Identify the target molecules in Chart 1 that will be formed as racemic mixtures. Explain why a racemic mixture is formed in some cases, but not in others.

3. Propose methods for synthesizing the target molecules in Chart 2 from acetylene.

4. Identify all of the target molecules in Chart 2 that will be formed as racemic mixtures.

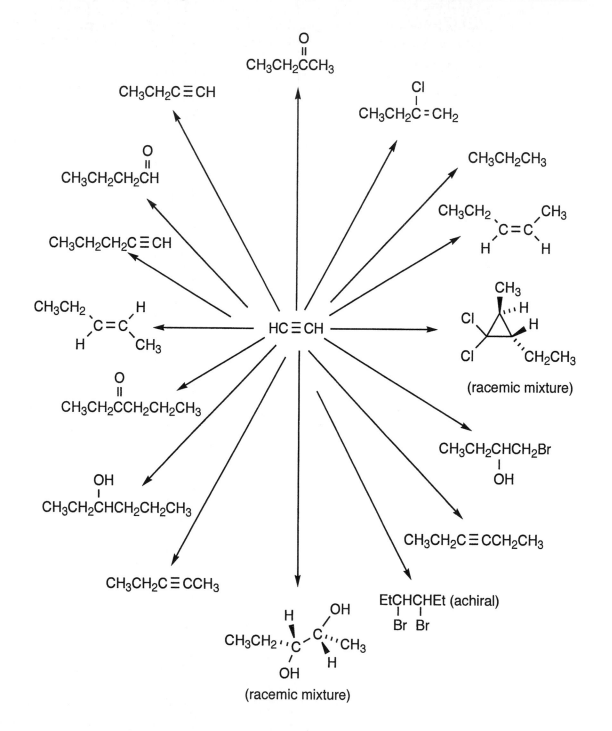

Chart 2

5.   Propose synthetic plans to carry out the transformations that follow. Specify necessary reagents and reaction conditions. Each transformation requires at least two steps. Start by brainstorming the problem to diagnose the major chemical challenge (e.g., make a C-C bond, control stereochemistry, control regiochemistry, etc.). Then, "back the problem up" by asking,

   a. What could be the penultimate (next-to-last) compound in the synthetic sequence?
   b. How could you convert the penultimate compound to the ultimate compound?
   c. If you cannot answer (b), then recycle to (a) and try another precursor to the final product.
   d. When you get a good final step, figure out how to make the penultimate compound, following the same logic until you get back to the starting material.

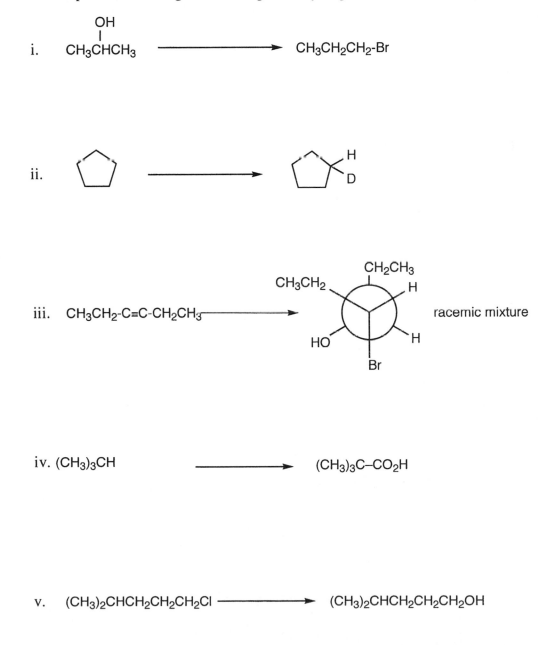

i.   CH₃CHCH₃ (with OH on middle carbon)   ⟶   CH₃CH₂CH₂-Br

ii.   (cyclopentane) ⟶ (cyclopentane with H and D)

iii.   CH₃CH₂-C≡C-CH₂CH₃ ⟶ (Newman projection: CH₃CH₂, CH₂CH₃, H, HO, H, Br)   racemic mixture

iv.   (CH₃)₃CH   ⟶   (CH₃)₃C–CO₂H

v.   (CH₃)₂CHCH₂CH₂CH₂Cl   ⟶   (CH₃)₂CHCH₂CH₂CH₂OH

Reflection: Work with your colleagues to construct a stepwise flowchart for solving a synthesis problem.

WORKSHOP 11

Chirality

Purpose: There are marvelous consequences of the tetrahedral geometry of the tetrasubstituted carbon and the three-dimensional structure of molecules. In this Workshop, you first deal with the problem of representing 3-D molecules in 2-D space. As part of that problem, you will expand your ideas about isomers and learn to distinguish and name different kinds of stereoisomers. By coupling ideas about stereochemistry to reaction chemistry, you should be able to predict or deduce the stereochemical characteristics of various reactions. You should also be able to explain how stereoisomers can be distinguished and separated.

Expectations: You should review and be prepared for an initial discussion about the meaning of the following key words and concepts:

   a.   chirality, chiral, achiral
   b.   optical activity, specific rotation
   c.   absolute configuration (R/S)
   d.   enantiomers, diastereomers, racemic, *meso*
   e.   *syn* and *anti* addition

1.   Identify all the stereocenters in the following compounds.

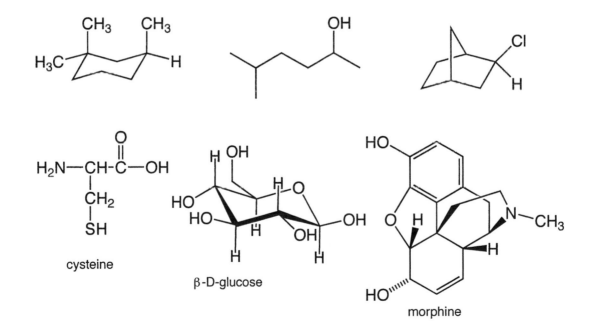

cysteine

β-D-glucose

morphine

69

2.   a.   Determine the relationship between the compounds in each of the pairs that follow. They may be **identical**, **constitutional isomers**, **enantiomers**, **diastereomers**, or **none of these**.

To clarify your understanding of these terms, construct a logic tree for classifying isomers.

b.   Identify the optically active (chiral) compounds and any *meso* compounds.

c.   Assign a **configuration (R or S)** to all stereocenters. Make molecular models of these compounds to confirm your assignments. Since rotation around σ bonds cannot change a molecule's constitution or configuration, you may change the conformations of these molecules.

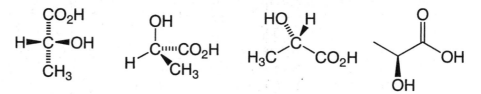

Note: among four structures, there are six pairwise relationships.

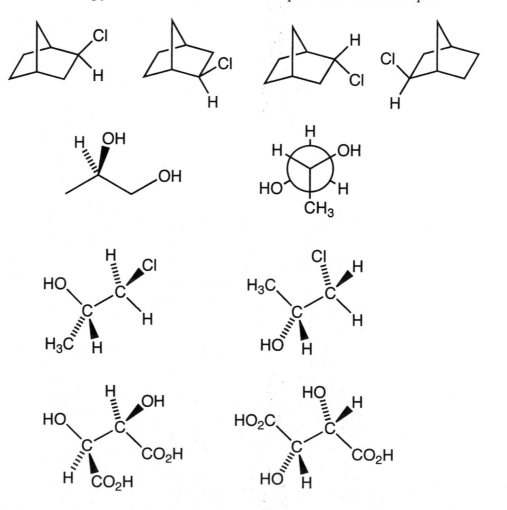

Midway Reflection: Before moving on to the rest of the problems in this Workshop, reflect on the strategies you used to determine the relationships between the various pairs of compounds. There are at least two different ways in which people carry out mental rotation tasks. Decide which ones you used, and then compare the approaches used by your classmates. Try methods you may not have already tried.

Object Rotation: One way to solve the problems is by imagining one figure rotating into correspondence with another and checking to see if they match. Sometimes it is possible to rotate the whole object at once (whole-object rotation). At other times it is necessary to break the object into pieces and rotate the parts one at a time, checking for a match at each step (piecewise rotation).

Object Inspection: Object inspection (e.g., inspection of the front of an object) involves analyzing the spatial relations in an object from a given viewpoint. One then adopts the same viewpoint with another object and checks whether the same left–right and top–bottom spatial relations hold (e.g., R or S designations could fall into this category if you imagine yourself viewing the stereocenter from a point opposite the lowest-priority group). In object inspection, the objects are not rotated or brought into correspondence with each other; instead, one imagines oneself moving around the object to achieve the same point of view with both.

3.   For each of the reactions that follow, give a careful representation of the structure of the product, and predict whether the product will be optically active, a racemic mixture, or achiral. **Explain your choice.**

a.   (+)–2–chlorobutane $\xrightarrow[\text{in CCl}_4]{\text{Br}_2,\ h\nu}$ 2–bromo–2–chlorobutane

b.   (+)–2–chlorobutane $\xrightarrow[\text{in CCl}_4]{\text{SO}_2\text{Cl}_2,\ h\nu}$ 1,2–dichlorobutane

Several other products are formed. This product is separated by gas chromatography and collected for analysis.

c.   (+)–2–chlorobutane $\xrightarrow[\text{in EtOH}]{\text{KOH}}$ $\xrightarrow[\text{ether}]{\text{HBr in}}$

d.   *cis*–2–butene $\xrightarrow[\text{in CCl}_4]{\text{Cl}_2}$

e.   (+)–1,3–dimethylcyclopentene $\xrightarrow[\text{in CCl4}]{\text{NBS}\ h\nu}$

f.   (+)-3-methylcyclopentene $\xrightarrow{\text{PtO}_2\ \text{H}_2}$

g.   *cis*–2–hexene $\xrightarrow{\begin{array}{l}1.\ \text{OsO}_4\\2.\ \text{NaHSO}_3,\ \text{H}_2\text{O}\end{array}}$

4. Alkenes react with peroxycarboxylic acids (RCO3H) to give three-membered ring cyclic ethers called epoxides. For example, 4-octene reacts with peroxyacids to give 4,5-epoxyoctane.

$$CH_3CH_2CH_2CH=CHCH_2CH_2CH_3 \xrightarrow{\;RCO_3H\;} CH_3CH_2CH_2CH-CHCH_2CH_2CH_3$$

4,5-epoxyoctane

Analyze Parts A and B as **OBSERVATION–DEDUCTION** problems about reaction mechanisms.

A. This reaction is performed on *cis*-4-octene. The 4,5-epoxyoctane that is formed **cannot** be separated into optically active molecules.

1. What is the structure of the 4,5-epoxyoctane formed from *cis*-4-octene?

2. What is the stereochemical path of the addition of the oxygen to the double bond (i.e., are the bonds to the oxygen formed *syn* or *anti* with respect to the double bond)?

4.   (continued)

B.   The 4,5-epoxyoctane from *cis*-4-octene is reacted with H_2O in the presence of an acid catalyst. The epoxide ring is opened to give a diol, 4,5-dihydroxyoctane.

$$CH_3CH_2CH_2\overset{\displaystyle \overset{O}{\triangle}}{CH-CH}CH_2CH_2CH_3 \xrightarrow[\ H_2O\ ]{H_3\overset{+}{O}\ cat} CH_3CH_2CH_2\overset{\displaystyle OH}{\underset{|}{CH}}-\overset{\displaystyle OH}{\underset{|}{CH}}CH_2CH_2CH_3$$

from *cis*-4-octene

4,5-dihydroxyoctane

This 4,5-octanediol **can** be separated into two optically active forms, namely, (+)-4,5- octanediol and (-)-4,5- octanediol.

1.   What is the structure of the 4,5- octanediol formed from *cis*-4-octene?

2.   What is the stereochemical path of the addition of water to the epoxide (i.e., is the new C-O bond formed *syn* or *anti* to the C-O bond of the epoxide?)

5. Amines are sufficiently basic to react with alkyl bromides to effect the elimination of HBr.

$$e.g., \quad CH_3CHCH_3 + R_3N \longrightarrow R_3NH^+Br^- + CH_3CH=CH_2$$
$$\underset{|}{Br}$$

Strychnine is a basic amine, isolated from the beans of *strychnos ignatii Berg.* Strychnine is optically active.

A 1,2-dibromocyclopentane was prepared by reacting cyclopentene with Br_2. This 1,2-dibromocyclopentane, which was optically inactive, was reacted with less than one equivalent of strychrine. After the reaction was complete, the unreacted 1,2-dibromocyclopentane was recovered and found to be optically active.

a. Give a structure for the optically active 1,2-dibromocyclopentane.

b. Explain, from first principles, how the unreacted 1,2-dibromocyclopentane became optically active.

c. Give a reaction-energy diagram that illustrates your explanation.

Reflection:

a. Identify at least two ideas about stereochemistry which you brought into this workshop that were incorrect or incomplete. Explain how your ideas changed as a result of the workshop. First develop your individual responses, then discuss them as a group, and then write out what you gained from the workshop.

b. Write a short paragraph that explains the differences between enantiomers and diasteromers and why it is useful to recognize the differences.

WORKSHOP 12

Nucleophilic Substitution Reactions

Purpose: The nucleophilic substitution reactions of alkyl halides and related compounds are one of the great proving grounds for developing and testing ideas about the relationships between structure and reactivity. The purpose of this Workshop is to understand the rates and stereochemistry (observations) and reaction mechanisms (deductions) of nucleophilic substitution reactions in terms of the experimental variables, such as the structures of the alkyl groups, the nature and concentration of the nucleophile, the nature of the leaving group (nucleofuge), and the nature of the solvent. The Workshop should also give you a good sense of the thought processes that lead to mechanistic hypotheses about these interesting reactions.

Expectations: Prior to this Workshop, review acid–base and electrophile–nucleophile concepts, steric effects (van de Waals repulsions), stereochemistry (enantiomers and diastereomers), carbocation stabilities, resonance, and S_N2/S_N1 mechanisms.

1. Discuss how the three experiments that follow are related to our general understanding of the mechanism of *biomolecular* nucleophilic substitution reactions (i.e., analyze each of the experiments as an **OBSERVATION–DEDUCTION** problem about the reaction mechanism). Building molecular models will help.

    a. Iodide ion is a good nucleophile and sodium iodide is quite soluble in acetone. By contrast, sodium chloride and sodium bromide have low solubilities in acetone. As a result, the reaction of alkyl bromides and alkyl chlorides with NaI/acetone can serve as a *simple* test reaction:

$$R\text{-}Br + Na^+ + I^- \xrightarrow{\text{acetone}} R\text{-}I + NaBr \downarrow \quad \text{(precipitate, solution becomes cloudy)}$$

   The rates of these reactions depend on the concentration of I^- and the concentration of the alkyl halides. The observed reactivity order of the following alkyl bromides with NaI/acetone is

$$CH_3(CH_2)_3CH_2Br > (CH_3CH_2)_2CHBr > (CH_3)_3C\text{–}CH_2Br > (CH_3CH_2)_3CBr$$

   OBSERVATION   **DEDUCTION**

1.   (continued)

b.   The optical rotation of a solution of (+)-2-bromobutane in diethyl ether does not change with time.   However, when tetrabutylammonium bromide (a soluble source of bromide ion) is dissolved in this solution, the rotation decreases slowly with time to zero. The NMR spectrum of the solution, however, does not change.

OBSERVATION   **DEDUCTION**

Construct a graph of the mole fraction of (+)-2-bromobutane and (−)-2-bromobutane as a function of time. Why does the rotation decrease to zero? Why does it not become negative?

1.   (continued)

c.   Consider the following observation:

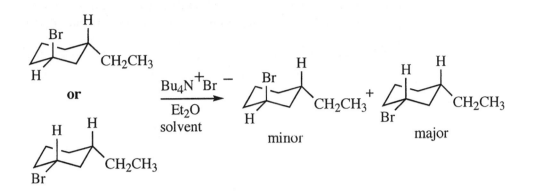

Two other experiments showed that exactly the same mixture is obtained from either pure starting material and that the rate of isomerization depends on [$Bu_4N^+Br^-$].

OBSERVATION   **DEDUCTION**

d.   On the basis of your **OBSERVATION–DEDUCTION** analyses, propose a generalized mechanism for these bimolecular nucleophilic substitutions. Explain how the observations in b and c are related to one another and how b and c are related to the observation in a.

2. Discuss how the following three experiments are related to our general understanding of the mechanism of *unimolecular* nucleophilic substitution reactions (i.e., analyze each of these experiments as an **OBSERVATION-DEDUCTION** problem about the reaction mechanism).

a. When 4–chloro–2–methyl–2–pentene reacts with acetic acid (solvent), two substitution products are formed, with the rearranged product predominating as shown in the diagram that follows. When small amounts of acetate ion are added to the reaction mixture, no increase in rate is observed. Propose a mechanism that accounts for these results. Explain clearly why two products are formed.

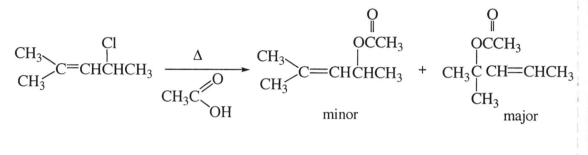

OBSERVATION   **DEDUCTION**

b. Low-molecular-weight alcohols dissolve in concentrated hydrochloric acid containing $ZnCl_2$. For some alcohols, a reaction takes place to form a new compound that is insoluble in concentrated $HCl/ZnCl_2$. Thus, a clear solution of the alcohol in the reagent turns cloudy. The rate of formation of this new product varies with structure:

$(CH_3)_3COH$ > $C_6H_5CH_2OH$ ≈ $CH_2=CHCH_2OH$ ≈ $(CH_3)_2CHOH$ > $CH_3CH_2CH_2OH$

What is the insoluble product? Give a balanced equation for the reaction that occurs.

OBSERVATION   **DEDUCTION**

2.   (continued)

c.   Analyze the following reaction:

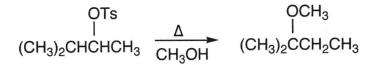

OBSERVATION   **DEDUCTION**

Construct a reaction-coordinate diagram that shows the relative energies of reactants, products, all proposed intermediates, and the transition states that interconnect them.

d.   Explain, on the basis of your **OBSERVATION–DEDUCTION** analyses, how the experimental observations in a, b, and c are related to one another. Identify the "common denominator" hypothesis that links your mechanistic proposals for reactions a, b, and c. Write a single, generalized mechanism for these unimolecular nucleophilic substitution reactions.

3.　　Consider the following observation about the addition of HCl to 3,3-dimethyl-1-butene:

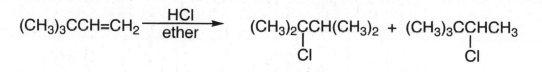

$$(CH_3)_3CCH{=}CH_2 \xrightarrow[\text{ether}]{\text{HCl}} (CH_3)_2\underset{\underset{Cl}{|}}{C}CH(CH_3)_2 \ + \ (CH_3)_3C\underset{\underset{Cl}{|}}{C}HCH_3$$

Review Workshop 7, and explain the relationship between this *addition* reaction in question 3 and the *nucleophilic substitution* reactions in question 2a.

4.　　Discuss with your classmates how molecular orbital ideas rationalize the stereochemistry of bimolecular nucleophilic substitution in Problems 1b and c and the two products observed in Problem 2a.

Reflection: We learn by making connections between otherwise isolated facts. This is, in part, the power of a reaction mechanism. Collaborate with your teammates to make a list of the new connections that you linked up as you worked through this Workshop.

WORKSHOP 13

Elimination Reactions

Purpose: Figuring out the ways that molecules react is the province of the mechanistic chemist. By understanding the mechanisms of reactions, the chemist can manipulate the experimental variables to favor one product over another. This element of control is the province of the synthetic organic chemist. This Workshop continues to explore how we know how molecules react and how this knowledge leads to control of the products.

Expectations: Prior to this Workshop, review acid–base and electrophile–nucleophile concepts, steric effects (van der Waals repulsions), stereochemistry (enantiomers and diastereomers), cyclohexane conformational analysis, carbocation stabilities, resonance, solvent effects, and the E2/E1 and S_N2/S_N1 mechanisms (Workshop 12).

1. Discuss how the experiments in a, b, and c are related to our general understanding of the mechanism of bimolecular elimination reactions (i.e., analyze these experiments as an **OBSERVATION–DEDUCTION** problem about reaction mechanisms.

   a. Even though the following reaction can give two stereoisomeric elimination products, only one is formed:

   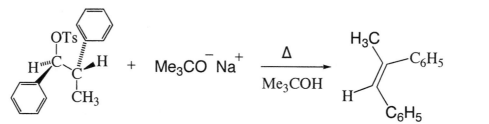

   b. The diastereomer of the tosylate shown in the preceding reaction also reacts with potassium *t*-butoxide to give an alkene that is diastereomeric to that obtained in the original reaction. Give the structure of the diastereomeric starting material and the structure for this alkene. Explain clearly what you can conclude from the observation that there are no crossover products.

   OBSERVATION   **DEDUCTION**

1. (continued)

    c. Give the products for the following two reactions:

        trans–2–methylcyclohexyltosylate $\xrightarrow[\text{Me}_3\text{COH}]{\text{Me}_3\text{CO}^-\text{Na}^+,\ \Delta}$ A

        cis–2–methylcyclohexyltosylate $\xrightarrow[\text{Me}_3\text{COH}]{\text{Me}_3\text{CO}^-\text{Na}^+,\ \Delta}$ A + B

        OBSERVATION   **DEDUCTION**

                A   **B**

    d. Propose a transition state for these bimolecular elimination reactions that is consistent with the experimental **OBSERVATIONS** and your **DEDUCTIONS** just given. Be sure to explain how the observations in part c are related to those in parts a and b.

2.   In parts 2a–e, discuss how changing the reaction conditions (e.g., the nature of the nucleophile or base and the solvent) leads to changes in reaction mechanisms and, therefore, to changes in reaction products.

a.   When the optically active tosylate **A** was reacted with $CH_3S^-Na^+$ in CH_3OH, the reaction was observed to be first order in tosylate and first order in nucleophile, and the substitution product shown was formed almost exclusively. Give a mechanism for this reaction and predict the stereochemistry of the product.

b.   When the solvent in part a was changed from methanol to dimethylformamide (DMF, $HCON(CH_3)_2$), the same product was formed, but at a much faster rate. Explain.

2.   (continued)

c.   When **A** was reacted with $CH_3O^-Na^+$ in CH_3OH, the reaction was still kinetically second order. However, the major product was an alkene. Account for the change in products as the reagents are changed from CH_3SNa/CH_3OH (part a) to CH_3ONa/CH_3OH (part b). Also, predict the stereochemistry of the methyl ether substitution product.

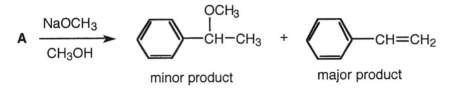

d.   How would you change reagents or reaction conditions so that **A** would give the alkene ($C_6H_5CH=CH_2$) almost exclusively?

e.   When **A** was refluxed in methanol (no added CH_3ONa), both substitution and elimination products were formed as in part (c). However, substitution predominates, and the stereochemical result was different from that observed for part (c). Predict the stereochemistry of the substitution product, and explain why these results are observed from the reaction of **A** in methanol.

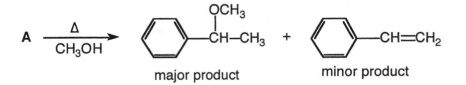

3. a. The reaction of 3-methyl-2-butanol with HBr in THF gives a mixture of 2-bromo-2-methylbutane (major) and 2-bromo-3-methylbutane (minor). The product ratio, [2-bromo-3-methylbutane]/[2-bromo-2-methylbutane], can be increased by an excess of NaBr. Give a complete mechanism that is in accord with all of these experimental observations.

OBSERVATION   **DEDUCTION**

b. The reaction of 3-methyl-2-butanol with hot H_3PO_4 gives three elimination products in yields of 65%, 33%, and 3%.

  i. Predict the structure of the products.

  ii. Give a complete mechanism for the formation of these three products.

  iii. Explain why substitution products form in part a, but are not formed in this case?

4.   On the following graphs, plot the qualitative change in reaction rate for $S_N1/E1$, S_N2, and E2 reactions as a function of the structure of the alkyl group (Graph 1) and the concentration of the nucleophile or base (Graph 2).

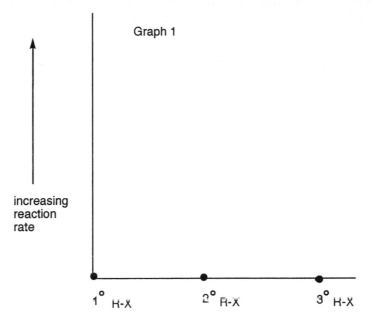

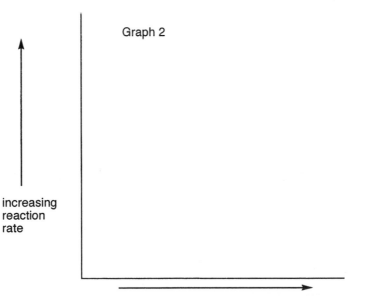

Reflection: Refer back to Workshops 7 and 8 to identify the connections among the reactions of alkenes and the reactions of alcohols in Problem 3 in this Workshop. Work with your teammates to construct a map which identifies the "BIG IDEA" that unifies your understanding of these reactions.

WORKSHOP 14

Alkyl Halides and Alcohols: Synthesis

Purpose: Victor Grignard won the Nobel prize in chemistry in 1912 because of the broad synthetic utility of his (Grignard) reagents for making carbon–carbon bonds and preparing alcohols. From a different perspective, the alcohol functional group stands at the center of a network of functional group interchanges. As a result, the chemistry of alcohols is a ubiquitous tool for organic synthesis. In addition, the mechanisms of many of the reactions of alcohols (and their derivatives) follow familiar patterns for substitutions, eliminations, and additions. This Workshop will help you review and consolidate your knowledge and understanding of functional group interchanges, reactions, mechanisms, redox chemistry, and organometallic procedures for C-C bond formation. In turn, the knowledge gained is prerequisite to finding solutions to challenges of organic synthesis.

Expectations: You should come prepared with an understanding of the nature and the chemistry of the carbon–metal bond in organolithium and organomagnesium compounds. Simultaneously, you should understand the redox relationships of alcohols to aldehydes and ketones, as well as the polar reactions of the alcohol functional group and its derivatives.

1.   Work with your colleagues to make a "reaction map" that shows the interchanges of other functional groups into alcohols or alcohols into other functional groups. For each interchange, link the reactants and products by specifying the necessary conditions and reagents.

Start by making a list of all the functional groups you have studied. You might also rank those groups by the oxidation state of carbon in order to identify interchanges that require redox reagents. That same ranking will help you construct a hierarchical reaction map, arranged by redox level. Because some of the chemistry is different, you will want to distinguish primary, secondary, and tertiary alcohols at the center of your map. Finally, you may find it worthwhile to use self-adhesive notes for the different interchanges so that you can rearrange and refine the map as you see new relationships and connections emerge.

2.   Specify all of the possible retrosynthetic disconnections to carbonyl compounds and Grignard reagents for the following compounds:

a.

$$\underset{\text{OH}}{\overset{\quad}{\text{CH}}}$$

OH
|
CH
CH$_3$

2.　　(continued)

　　b.　　CH_3CH_2OH

c.　　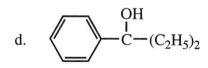

d.

$$\text{d.} \quad \underset{}{\bigcirc\!\!\!\!\bigcirc} \overset{\text{OH}}{\underset{|}{C}}-(C_2H_5)_2$$

3. Show how to carry out the synthetic transformations that follow. You may use methanol and ethanol as reactants, any inorganic reagents, and any necessary solvents. If more than one synthetic step is required, show the product for each step.

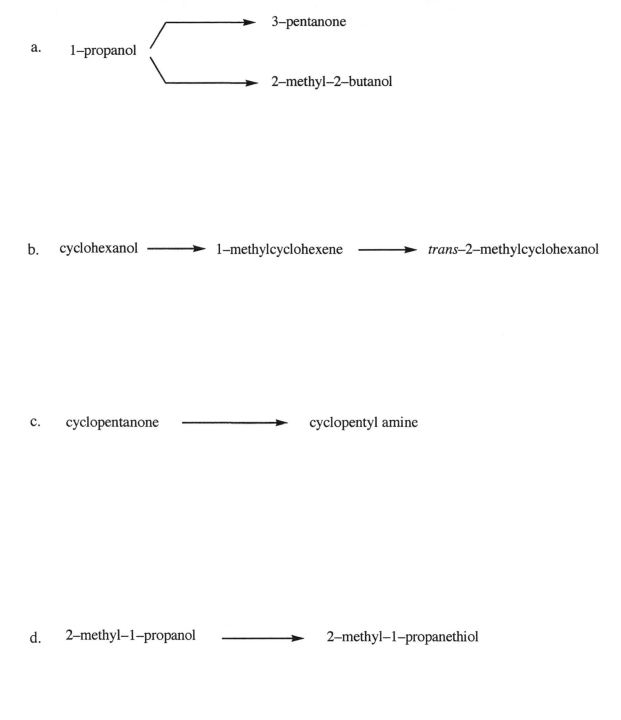

a. 1–propanol
    - 3–pentanone
    - 2–methyl–2–butanol

b. cyclohexanol ⟶ 1–methylcyclohexene ⟶ *trans*–2–methylcyclohexanol

c. cyclopentanone ⟶ cyclopentyl amine

d. 2–methyl–1–propanol ⟶ 2–methyl–1–propanethiol

3.   (continued)

e.   isopropyl bromide
      (only carbon source)

f.

g.   cyclopentane $\longrightarrow$ deuteriocyclopentane (C_5H_9D)

h.   cyclopentene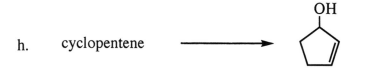

i.   1-chloropentane $\longrightarrow$ 1-heptanol

94

4.   Different reagents and conditions are used to convert tertiary alcohols or secondary alcohols to alkyl halides or alkenes.

Specify appropriate reagents for the following conversions, and explain, in mechanistic terms, why different reagents and conditions are required:

a.

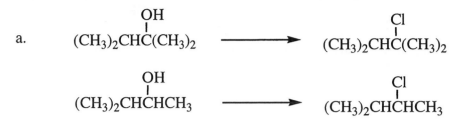

b.

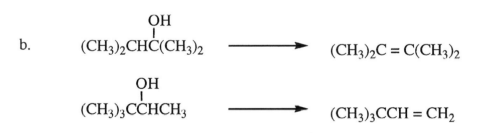

5.   Consider each of the reactions that follow. First, decide whether the reaction is net oxidation, net reduction, or neither. Second, specify an appropriate reagent to bring about the transformation. Third, write a balanced equation, and if the reaction is a redox reaction, identify the conjugated redox pairs, red-1/ox-1 and red-2/ox-2.

a.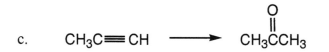

b.   And the reverse reaction?

c. $CH_3C \equiv CH \longrightarrow CH_3\overset{O}{\overset{\|}{C}}CH_3$

6. Thiols react with I_2 to give disulfides, which in turn react with Zn in acidic solution to give thiols. The cysteine–cytstine interconversion is an important biochemical example of this thiol–disulfide interconversion, albeit with different reagents.

$$RSH \underset{Zn, H_3O^+}{\overset{I_2}{\rightleftarrows}} RSSR$$

a. Balance these two reactions: thiol to disulfide and the reverse transformation of disulfide to thiol.

b. Analyze the preceding "atom-balanced" reactions to show the electron balance. Write the two redox half-cell reactions for each transformation. Be sure to label the half-cell reactions to specify the oxidized and reduced partners.

Reflection: Brainstorm with your colleagues to identify "BIG IDEAS" (recurring themes) that came up in your review of the pathways for the preparation and reactions of alcohols.

WORKSHOP 15

Epoxides and Ethers

Purpose: The reactions of epoxides are interesting and useful. In contrast to simple ethers, epoxides react readily with nucleophiles. Since epoxides are usually prepared from the corresponding alkenes, this Workshop provides an opportunity to analyze redox relationships. You should be able to (1) identify reactions that require redox reagents and (2) predict the stereochemistry and regiochemistry of additions to epoxides.

Expectations: You should review and be prepared for an initial discussion about the meaning of the following key words and concepts: oxidation–reduction, oxidation state, stereochemistry of addition, regiochemistry of addition, and acid and base catalysis.

1. Identify the oxidation state of each carbon in the functional groups that follow. For each pairwise transformation, describe it as an oxidation, a reduction, or neither. (Considering the forward and reverse reactions, there are 12 combinations.) For each transformation, indicate a reagent or sequence of reactions that would be appropriate to accomplish the transformation. Write a balanced redox equation.

$$CH_2 = CH_2 \qquad CH_3CH_2OH \qquad HOCH_2CH_2OH \qquad \underset{H_2C-CH_2}{\overset{O}{\triangle}}$$

2. Each of the synthetic reactions that follow failed. Explain what went wrong and indicate which products, if any, would be formed. Offer alternative synthetic approaches.

a. ⬡–$CH_2O^-\ Na^+$ + ⬡–Br ⟶ ⬡–CH_2O–⬡

b. $CH_3CH_2O^-\ Na^+$ + $H_3C–\overset{\overset{CH_3}{|}}{\underset{\underset{CH_3}{|}}{C}}–Br$ ⟶ $CH_3CH_2O–\overset{\overset{CH_3}{|}}{\underset{\underset{CH_3}{|}}{C}}–CH_3$

c. [epoxide structure] + HBr ⟶ $BrCH_2(CH_2)_3\overset{\overset{OH}{|}}{C}(CH_3)_2$

3. Propose a reasonable mechanism for the reactions that follow. Use the arrow formalism to show the bond-making and bond-breaking processes.

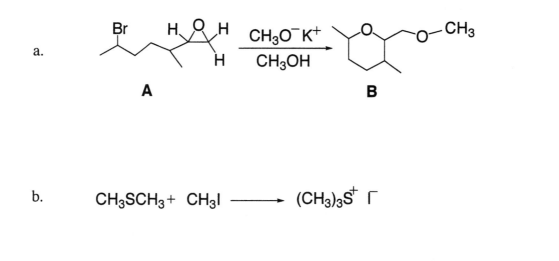

a.

A   B

b. CH_3SCH_3 + CH_3I ⟶ $(CH_3)_3S^+\ I^-$

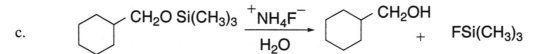

c.

4. The mechanism of the ring opening of styrene oxide was studied under various experimental conditions (*J.Org. Chem.*, **1994**, *59*, 1638–1641.)

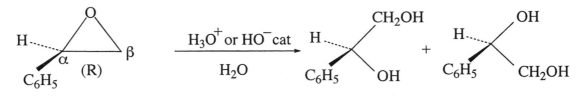

The glycols were esterified by reaction with the (+) enantiomer of the acid chloride (MTPA). The resulting mixture of diesters was

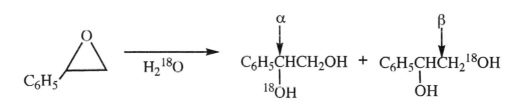

The products of the reaction were analyzed by NMR. The benzylic hydrogens of the resulting esters gave distinct multiplets centered at $\delta = 6.17$ and $\delta = 6.28$. The former corresponded to an authentic sample of the diester of (+)MTPA and (R)-glycol. Simple integration of the multiplets gave the (R)/(S) composition of the glycols.

In a separate experiment, the percent substitution at the α- and β-carbons was determined by isotopic labeling with $H_2{}^{18}O$:

Catalyst	%α	%β	Glycol	
			% (S)-	% (R)-
HO⁻	50	50	52	48
None	95	5	93	7
H_3O^+	100	0	67	33

4.   (continued)

a.   Analyze the data in the table, using the **OBSERVATION–DEDUCTION** methodology to come to conclusions about the mechanisms of the substitution reactions under the three different experimental conditions. Be sure that your mechanisms specify the reacting nucleophile and electrophile, the regiochemistry, and the stereochemistry. It will help to construct multiple hypotheses about the mechanism and to compare your predictions with experiment. You will need to be alert to the possibility of competing reaction paths. Start with the base-catalyzed reaction, and consider which labeled species are present when sodium hydroxide is dissolved in water that is isotopically enriched with ^{18}O.

base catalyzed

uncatalyzed

acid catalyzed

4. (continued)

    b.   What do the experiments tell you about the stereochemistry of the base-catalyzed reaction at the β-carbon? Work with your colleagues to design an experiment to reveal the stereochemistry of β-substitution.

    c.   How do you think the position of the ^{18}O label could be determined?

5. a. Draw a Newman projection of the epoxide prepared from (E)-2-butene by reaction with peracetic acid.

b. Given what you know about the mechanism of ring opening of epoxides in basic solution, draw a Newman projection of the transition state for the reaction of the epoxide from (E)-2-butene in aqueous base. Be sure to carefully represent the stereochemical relationship of the bond being made to the bond being broken in the transition state.

c. Give a Newman projection of the immediate product of the reaction, before any rotation around the 2,3-bond. Describe the stereochemical relationship of the two C-O bonds in the product.

5.   (continued)

    d.   Now, consider the ring opening of cis-1,2-epoxycyclohexane in basic aqueous solution. Using what you learned in parts a, b, and c of this problem, predict the immediate product of the ring-opening reaction. Give a careful chair representation and a name of this product.

    e.   Consider whether the product in part d is the more stable isomer. If not, give a careful chair representation and a name for the more stable isomer.

Reflection:

a.   Explain how you might use the redox analyses in Problem 1.

b.   Discuss the value of making multiple hypotheses in Problem 4.

WORKSHOP 16

Conjugated Systems

Purpose: The functional group concept in organic chemistry is one of the great classification schemes in science. It is especially powerful because it is based on the important insight that properties follow from structure; alcohols share chemical and physical properties because they share the structure of the alcohol functional group. Implicit in the functional group classification scheme is the idea that the structural units are isolated and independent. A molecule can have two (or more) noninteracting functional groups, and the properties are typically the sum of the properties of those groups. 1,4-Pentadiene is a good example of an independent functional group because an intervening $-CH_2-$ unit isolates the two double bonds.

In contrast to molecules with noninteracting functional groups are molecules in which the functional groups *are* close enough to interact. The result is a new structural unit with its own unique structure and, therefore, its own unique properties. Such arrangements are called conjugated systems; the functional groups are conjugated ("married") to each other. 1,3-Pentadiene is a good example of a conjugated system. As we will see, the properties of the conjugated double bonds are different from the properties of the isolated double bonds in 1,4-pentadiene.

This Workshop should provide a good understanding of the origins of the unique properties of conjugated systems.

Expectations: You should review and be prepared to use the following observations and concepts: reactivity of allylic systems; π-molecular orbital diagrams for 3 and 4 adjacent *p*-orbitals; ultraviolet absorption spectroscopy; heats of formation; 1,2 and 1,4 addition to conjugated dienes; and relative reactivity of conjugated and nonconjugated dienes.

1. Consider the following experimental observations.

    a. 1,3-cyclohexadiene

    $$\xrightarrow[\text{in ether}]{\text{DCl}} C_6H_8DCl$$

    1,4-cyclohexadiene

    Relative rate of reaction: 1,3-cyclohexadiene > 1,4 cyclohexadiene

    b. 1,3-cyclohexadiene $\xrightarrow[\text{in ether}]{\text{DCl}}$

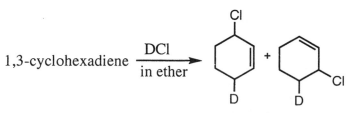

    unequal amounts

1.   (continued)

c.   cyclohexene $\xrightarrow[\text{CCl}_4,\ h\nu]{\text{N-bromosuccinimide}}$ 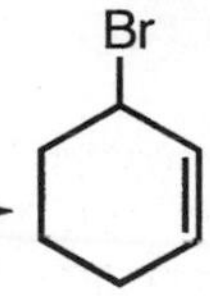   (the 1-bromo- and 4-bromo-derivatives are not formed)

d. 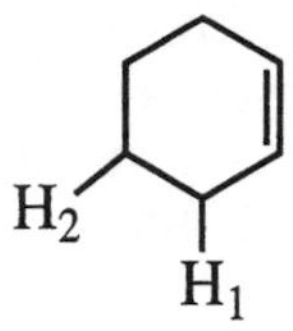 $pK_a\ H_1 \sim 46$

$pK_a\ H_2 \sim 52$

(1)   Give mechanisms for each of the chemical reactions a through d. Be careful to identify the reactive intermediate in each case.

1.   (continued)

(2)   Construct an energy-level diagram and orbital pictures for the three π molecular orbitals that result from the interactions of three adjacent (conjugated) p-orbitals.

(3)   Use the molecular orbital diagram to specify the electron configuration of the π system of each reactive intermediate in observations a through d.

1.　　(continued)

    (4)　　Give reaction-energy diagrams that describe each of the experimental observations. Explain how the molecular orbital descriptions and the reaction-energy diagrams rationalize those observations.

2. The heats of formation of 3-chloro-1-butene and 4-chloro-1-butene from the elements are –45.6 kJ/mol and –28.9 kJ/mol, respectively. The rate of formation of chloride ion from
3-chloro-1-butene in aqueous ethanol at 50° is approximately 10^3 faster than from 4-chloro-1-butene. Give a reaction-energy diagram that summarizes these observations. Explain clearly the origin of the energy differences that are responsible for the observed rate difference.

3. Consider the conformational changes in 1,3-butadiene that correspond to rotation around the 2,3 bond.

   a. Identify conformations that correspond to (two) energy minima and (one) energy maximum.

   Give Newman projection representations of these conformations; look down the 2,3 bond.

   b. Which of the two energy minima predominates at equilibrium? Explain your choice.

3.   (continued)

    c.   In butane, the eclipsed conformations correspond to energy maxima. Explain why 1,3-butadiene is different.

    d.   Give a reaction-energy diagram corresponding to rotation around the 2,3 bond in 1,3-butadiene. Assign structures to the maximum and the minima.

    e.   Barriers to rotation around C-C bonds are as follows:

ethane	2.9 kcal/mol
1,3 butadiene (the 2-3 bond)	3.9 kcal/mol
(E)-2-butene (the 2-3 bond)	~60 kcal/mol

Identify the origins of these different barriers.

4. Construct a π-molecular orbital diagram corresponding to 1,3-butadiene from the π and π* orbitals of two ethylene molecules.

    a. Start by interacting two *p*-orbitals to construct the π and π* orbitals of the ethylenes. Represent the relative energies of these orbitals qualitatively on the right and left sides of a diagram; set the energy of noninteracting *p*-orbitals as zero on this diagram.

    b. Since orbitals that are closest in energy interact most strongly, construct the π molecular orbitals of 1,3-butadiene (four adjacent conjugated *p*-orbitals) by interacting the π orbitals of the two ethylenes in phase and out of phase. Represent the relative energies of the resulting π molecular orbitals (π_1 & π_2) on your diagram.

    c. Construct the π* molecular orbitals in similar fashion by in-phase and out-of-phase interactions of the π* orbitals of the two ethylenes. Represent the energies of these orbitals (π_1* and π_2*) on your diagram.

4.   (continued)

      d.   Given that the absorption of a photon of appropriate energy excites an electron from the HOMO to the LUMO, use your M.O. diagrams to explain why ethylene absorbs ultraviolet light at 167 nm and 1-3-butadiene absorbs at 217 nm.

      e.   Consider the structures of ethylene acetone and 3-butene-2-one (methyl vinyl ketone), as well as the corresponding $\pi \rightarrow \pi^*$ absorptions. Reason by analogy to explain why methyl vinyl ketone absorbs a longer wavelength than either of the two constituent functional groups.

$$CH_2{=}CH_2 \quad 167 \text{ nm} \qquad\qquad CH_3\underset{O}{\overset{O}{C}}CH{=}CH_2 \quad 219 \text{ nm}$$

$$CH_3\overset{O}{\overset{\|}{C}}CH_3 \quad 187 \text{ nm}$$

      f.   1,2-Butadiene (methylallene) absorbs in the UV at 178 nm. Explain how our understanding of the structures of 1,3- and 1,2-butadiene are consistent with the observed UV absorptions.

Reflection: Look up the structure of rhodopsin (the light-sensitive pigment in your eyes) in your textbook. Explain how a low-energy, long-wavelength photon of visible light leads to isomerization of the C-11 double bond.

REVIEW

Purpose: Studying for a final exam can be numbing if your strategy is simply to go over (repeat) the semester's work. A better approach is to find new ways to organize the material to make new connections. This is much more than a technique to stay awake; we learn by making connections and perceiving relationships. The purpose of this Workshop is to revisit "BIG IDEAS" and fundamental thinking skills, in ways that move them from specific contexts to general themes, from knowledge to ways of knowing.

Expectations: You will have the most fun and be most productive if you construct your own problems (and the corresponding answers). All of the problems in this Workshop can be extended in new directions with great rewards for you and your colleagues. Problems, like exams, help you identify gaps in your understanding. As you work through these problems and others, make notes for yourself about holes that need to be filled and concepts that need to be reviewed. Then be sure to follow up!

1. The reactions that follow occur by analogous mechanisms. Explain how the reactions within each group are related. It will help to identify the electrophile and nucleophile for each mechanistic step.

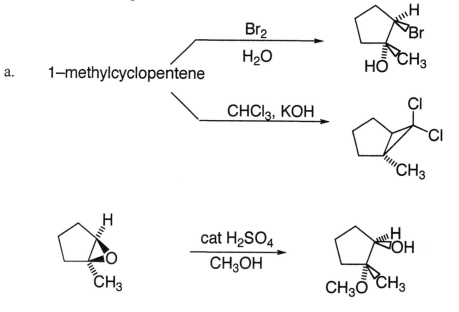

116

1.   (continued)

b.   methyl 2-methyl-2-propyl ether   HBr→   2-bromo-2-methylpropane + CH_3OH

2-methyl-2-propanol   HBr→   2-bromo-2-methylpropane

2-methylpropene   HBr→   2-bromo-2-methylpropane

c.

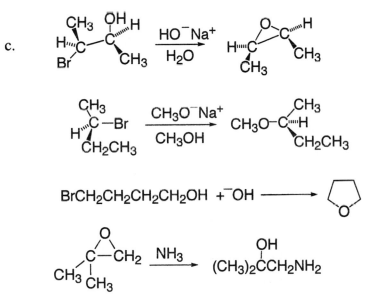

1.   (continued)

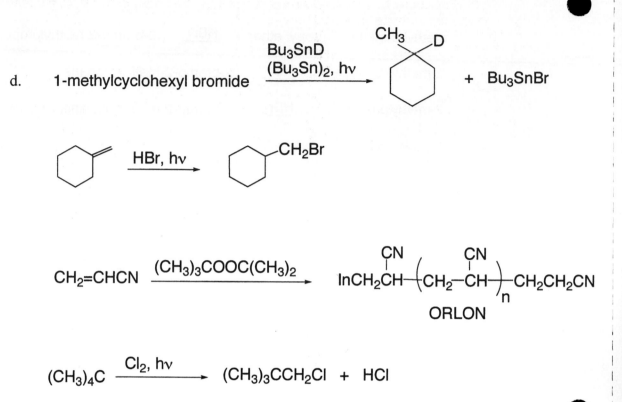

d.   1-methylcyclohexyl bromide $\xrightarrow[\text{(Bu}_3\text{Sn)}_2,\text{ h}\nu]{\text{Bu}_3\text{SnD}}$ + Bu$_3$SnBr

$\xrightarrow{\text{HBr, h}\nu}$

CH$_2$=CHCN $\xrightarrow{\text{(CH}_3)_3\text{COOC(CH}_3)_2}$ InCH$_2$CH$-$(CH$_2$$-CH)_n$$-CH_2CH_2$CN

ORLON

(CH$_3$)$_4$C $\xrightarrow{\text{Cl}_2,\text{ h}\nu}$ (CH$_3$)$_3$CCH$_2$Cl + HCl

2.   Work with your colleagues to carry out the indicated synthetic conversions in an efficient
       manner. Find as many other interconversions as you can. If more than one step is
       required, show the products for each step.

a.

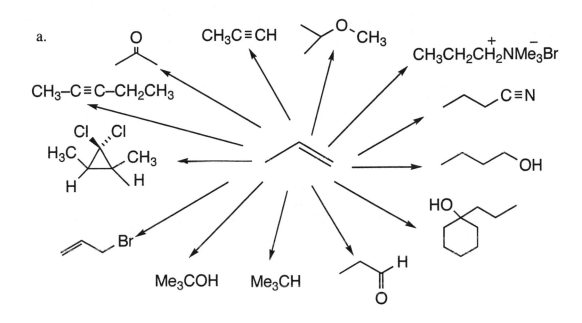

2. (continued)

b.

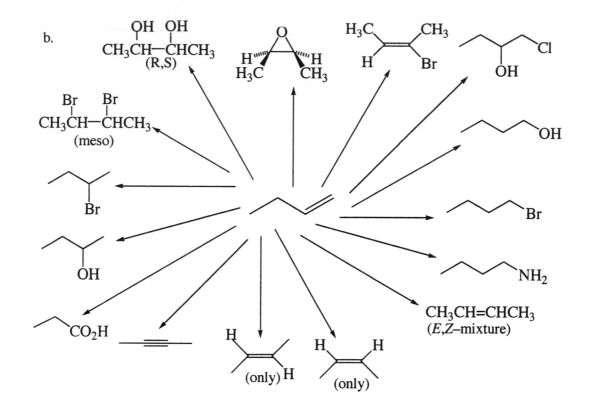

2.   (continued)

c.

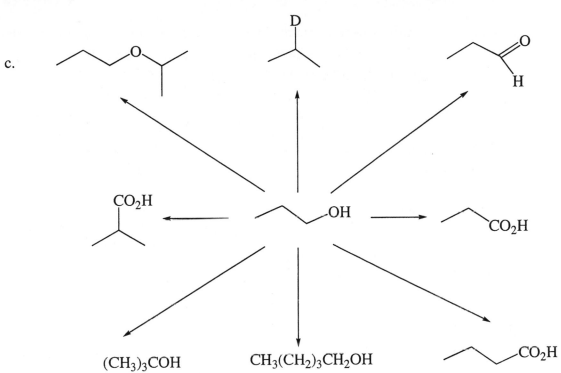

(CH₃)₃COH   CH₃(CH₂)₃CH₂OH

2.   (continued)

d.   Show how to carry out the indicated synthetic conversions in an efficient manner. Find as many other interconversions as you can. If more than one step is required, show the products for each step.

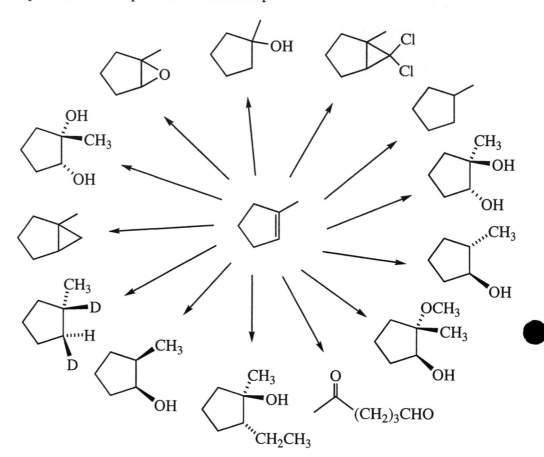

3. For each pair, circle the reaction that occurs at a *faster* rate. Justify your answer clearly, and identify the key structural feature that determines $\Delta\Delta G^{\ddagger}$. (*Hint*: What is the mechanism?) Work on this problem in pairwise fashion, taking turns to explain the following reaction pairs to each other:

a.   2–butyne + Br_2 $\xrightarrow{Et_2O}$ *trans*–2,3–dibromo–2–butene

   trans–2–butene + Br_2 $\xrightarrow{Et_2O}$ *meso*–2,3–dibromobutane

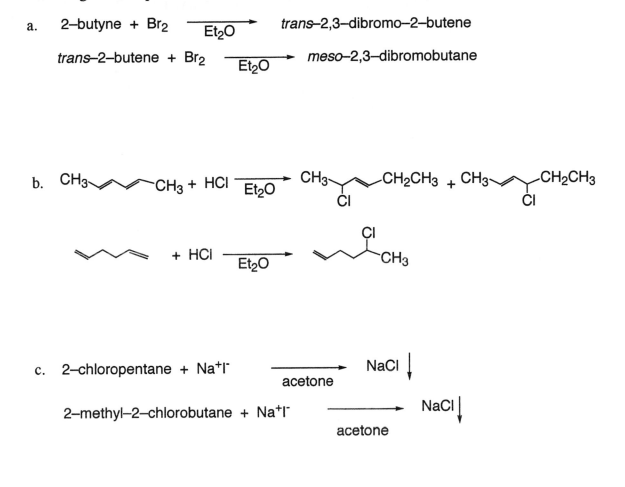

b.

c.   2–chloropentane + Na^+I^- $\xrightarrow{acetone}$ NaCl ↓

   2–methyl–2–chlorobutane + Na^+I^- $\xrightarrow{acetone}$ NaCl ↓

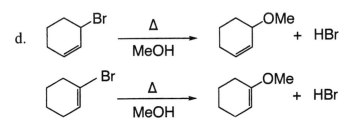

d.

3.   (continued)

e.   *trans*–2–methylcyclohexyl chloride $\xrightarrow[\text{(CH}_3)_3\text{COH}]{\text{(CH}_3)_3\text{CO}^-\text{K}^+}$ 1–methylcyclohexene

3–methylcyclohexene

f.   *t*–butanol $\xrightarrow[\text{HCl}]{\text{ZnCl}_2}$ *t*–butyl chloride

isopropanol $\xrightarrow[\text{HCl}]{\text{ZnCl}_2}$ *t*–isopropyl chloride

g.   1-butene $\xrightarrow[\text{Et}_2\text{O}]{\text{HBr}}$ 2-bromobutane

1,3-pentadiene $\xrightarrow[\text{Et}_2\text{O}]{\text{HBr}}$ 1-bromo-3-butene + 3-bromo-1-butene

3.   (continued)

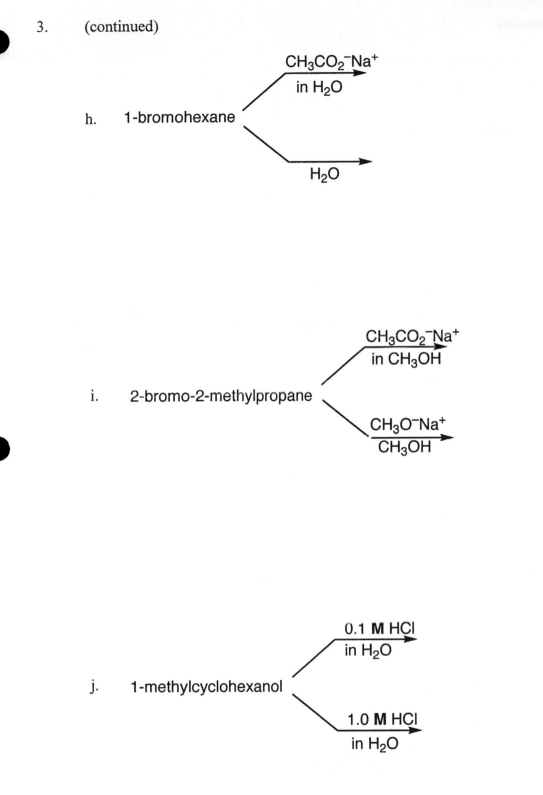

h.   1-bromohexane

$CH_3CO_2^-Na^+$
in H_2O

H_2O

i.   2-bromo-2-methylpropane

$CH_3CO_2^-Na^+$
in CH_3OH

$CH_3O^-Na^+$
CH_3OH

j.   1-methylcyclohexanol

0.1 **M** HCl
in H_2O

1.0 **M** HCl
in H_2O

3.   (continued)

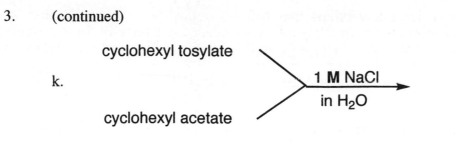

k.

cyclohexyl tosylate

cyclohexyl acetate

$\xrightarrow[\text{in } H_2O]{\text{1 M NaCl}}$

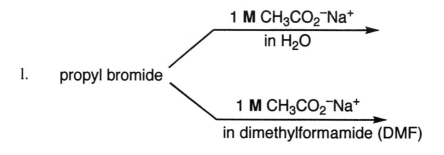

l.   propyl bromide

$\xrightarrow[\text{in } H_2O]{\text{1 M } CH_3CO_2{}^-Na^+}$

$\xrightarrow[\text{in dimethylformamide (DMF)}]{\text{1 M } CH_3CO_2{}^-Na^+}$

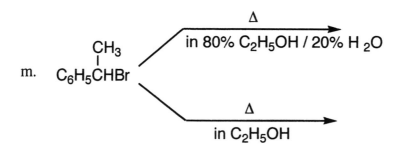

m. $C_6H_5\overset{\overset{\displaystyle CH_3}{|}}{C}HBr$

$\xrightarrow[\text{in 80\% } C_2H_5OH \text{ / 20\% } H_2O]{\Delta}$

$\xrightarrow[\text{in } C_2H_5OH]{\Delta}$

4. For each of the interconversions that follow, circle the compounds that would predominate at equilibrium. Justify your choice clearly, and identify the key structural feature that determines $\Delta G°$. Work in pairwise fashion, taking turns to explain your choices to each other.

a. $CH_3CH_2CH_2–MgBr + CH_3C{\equiv}CH \rightleftharpoons CH_3CH_2CH_3 + CH_3C{\equiv}C–MgBr$

b.
$$\overset{OH}{\underset{|}{CH_3CH=CCH_3}} \rightleftharpoons \overset{O}{\overset{\parallel}{CH_3CH_2CCH_3}}$$

c. $HC{\equiv}CH + OH^- \rightleftharpoons HC{\equiv}C^- + H_2O$

d. $CH_3\overset{+}{O}{\overset{H}{\underset{H}{\diagdown}}} + CH_3CH_2NH_2 \rightleftharpoons CH_3OH + CH_3CH_2\overset{+}{N}H_3$

127

4.   (continued)

e.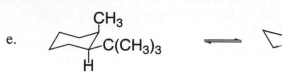

f.   2,3–dimethyl–2–butene $\xrightarrow{H_2SO_4}$ 2,3–dimethyl–1–butene

g. $CH_3Cl + CN^- \rightleftharpoons CH_3CN + Cl^-$

h.   cyclopropane $\rightleftharpoons$ propene

5.   Brainstorm with your colleagues to make lists showing all the examples you know of reactions that

   a.   form new carbon–carbon bonds;

   b.   are useful to control the stereochemistry of the product;

   c.   involve the oxidation of an organic molecule;

   d.   involve the reduction of an organic molecule;

   e.   cleave carbon–carbon bonds.

6.   For each pair of isomers that follows, suggest a simple, fast, qualitative laboratory procedure that would distinguish the members of the pairs. For example, alkanes and alkenes can be easily distinguished because alkenes react rapidly with permanganate ion in an aqueous base to give a black-brown precipitate of MnO_2. Alkanes do not react. You should consider both chemical and spectroscopic methods in each case. Be very specific about what you would do and what you expect to observe.

   a.   1-hexanol and *n*-pentyl methyl ether

   b.   1-hexyne and 2-hexyne

   c.   *n*-butanol and *tert*-butanol

6.   (continued)

     d.   1-bromopentane and 1-chloropentane

     e.   1-bromocyclohexene   and   3-bromocyclohexene

     f.   2,3-dimethyl-2-butene   and   cyclohexane

     g.   cyclohexanol   and   1-hexene-3-ol

7.   The curved-arrow formation is a powerful way to represent electron movement in chemical reactions, but it leaves open the question of where the electron goes. For each of the reactions that follow, identify the key orbitals (HOMO/LUMO) that help you understand the bond-making and bond-breaking processes. Start by identifying acids (electrophiles) and bases (nucleophiles) in the polar reactions; use the curved-arrow formalism to represent the reaction. Then, identify the HOMO and LUMO orbitals that are involved. Next, extend the ideas so that you can understand the bond-breaking processes in the last three electron transfer reactions. Finally, give products for each of the reactions.

$$H_2O + {}^+C(CH_3)_3 \longrightarrow$$

$$H_2O + H\overset{+}{N}H_3 \longrightarrow$$

$$CH_3Cl + H_2O \longrightarrow$$

$$(CH_3)_2C{=}CH_2 + Br_2 \longrightarrow$$

$$(CH_3)_3C{=}O + H_2O \longrightarrow$$

$$H_2O + C_3H_5^+ \text{ (allyl cation)} \longrightarrow$$

$$Mg + CH_3Cl \longrightarrow$$

$$Na + HC{\equiv}CH \longrightarrow$$

$$\textit{cis-}2\text{-butene} \underset{\longleftarrow}{\overset{h\nu}{\longrightarrow}} \textit{trans-}2\text{-butene}$$

8. Consider the following observations: On the one hand, 2-bromo-2-methylbutane reacts with $CH_3CH_2O^- K^+$ in CH_3CH_2OH at 25 °C to give a mixture of two products, 2-methyl-1-butene (30%) and 2-methyl-2-butene (70%). On the other hand, when either 2-methyl-1-butene or 2-methyl-2-butene is treated with a catalytic amount of concentrated sulfuric acid at 25 °C, the same mixture of the two butenes is obtained; 2-methyl-2-butene predominates (92%) in this mixture.

    a. Explain clearly why these two different procedures give different mixtures of 2-methyl-1-butene and 2-methyl-2-butene.

    b. Give a reaction-energy diagram that illustrates your explanation in part a. Be sure to label the important energy differences on your diagram so that they correlate with your explanation.

9. 3-Hexyne is reduced by $LiAlH_4$ at 138°C to give a 3-hexene that reacts with Br_2 in ether to give a 3,4-dibromohexane. This 3,4-dibromohexane **cannot** be separated into optically active enantiomers. Here is the sequence of reactions:

$$3\text{-hexyne} \xrightarrow[\text{in diglyme}]{LiAlH_4, 138°C} 3\text{-hexene} \xrightarrow[\text{in ether}]{Br_2} 3,4\text{-dibromohexane}$$

a. What is the structure of the 3,4-dibromohexane formed in this sequence of reactions? Explain your reasoning.

the 3,4-dibromohexane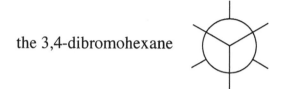

b. What is the structure of the 3-hexene formed in the reduction reaction? Explain your reasoning.

c. What is the stereochemistry of the reduction of the alkyne by $LiAlH_4$; that is, are the hydrogens added in *syn* or *anti* fashion? Explain your reasoning.

d. Propose a mechanism for the reaction of $LiAlH_4$ with alkynes. Use the experimental OBSERVATIONS to DEDUCE a mechanism.

10. (S)-2,3-dimethyl-1-pentene is reacted with $BH_3 \bullet THF$, followed by reaction of the intermediate borane with $H_2O_2/HO^-/H_2O$.

    a. Predict the "constitutional" structure of the product; that is, which constitutional isomer is formed?

    b. Predict the stereochemical structure of the product(s). Consider the following possibilities and explain the basis of your prediction:

        i. a single stereoisomer of the product;
        ii. a racemic mixture of two enantiomers of the product;
        iii. a mixture of two diastereomers of the product in unequal amounts;
        iv. a mixture of two diastereomers in equal amounts;
        v. a mixture of four stereoisomers of the product.

11.   Give structures for the lettered compounds.

A hydrocarbon, **K**, has the formula $C_{10}H_{18}$ and is optically inactive. **K** reacts with two equivalents of H_2 in acetic acid solvent in the presence of PtO_2 catalyst to give **L**. **K** gives only one product, **M**, when reacted either with $HgSO_4$ /H_2SO_4/H_2O or with $BH_3 \bullet THF$ followed by H_2O_2/^-OH/H_2O. Ozonolysis of **K** or reaction with $KMnO_4^-$ /H_3O^+/H_2O gives only **N**, $C_5H_{10}O_2$.   **N** is easily separated into optically active compounds (+)-**N** and (-)-**N**.

OBSERVATION   DEDUCTION

K =   L =

M =   N =

12. a. Construct a "reaction map" for carbocation intermediates. Build the map so that it summarizes all of the functional group transformations (reactants and products) that proceed by carbocation mechanisms.

b. Construct a "concept" map for the carbocation hypothesis. Build the map by providing linking words to describe the relationships among the central concept (carbocation) and subsidiary concepts. Start by ranking the subsidiary concepts in rough order of importance. This will provide the basis for constructing a hierarchical map.

Subsidiary Concepts

Structure
Heterolysis
Hyperconjugation
Bromonium Ion
Nucleophile
Empty Orbital
Rearrangement
sp^2
Rate-Determining Step
Allylic
Acid
Benzylic
Hammond Postulate

Electrophile
Trigonal Planar
Relative Stability
Reactivity
Energy Minimum
Mass Spectrometry
Acid
Delocalized
Product-Determining Step
Achiral
LUMO
Nonbonding Molecular Orbital
Competitive Reactions

WORKSHOP 17

Aromaticity

Purpose: As we have seen, organic chemistry is about the interplay of structure and reactivity. Workshop 16 introduced the idea that the special properties of allylic intermediates and dienes are a consequence of the special structures of conjugated systems. In this Workshop, we will extend this idea to include the remarkable properties of *cyclic*, continuously conjugated molecules and ions.

Expectations: Be prepared to explain, discuss, and use the following ideas: sp^2 hybridization; pK_a, molecular orbital, and resonance descriptions of the structure of benzene; Hückel's rule; aromatic; antiaromatic; nonaromatic; planarity; chemical shift; and ring current shielding and deshielding.

1. Predict whether each of the systems that follow is aromatic, antiaromatic, or nonaromatic. Assume that each system is planar, and justify your choice. Predict specific chemical or physical properties that would reveal the aromatic, antiaromatic, or nonaromatic character of the compound.

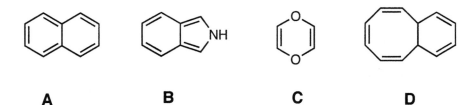

| A | B | C | D |

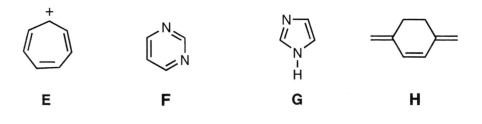

| E | F | G | H |

2.   Provide reasonable explanations for the following experimental observations:

a.   Neither of the following [10] annulenes shows any special stability. (It may help to make models.)

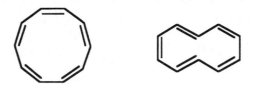

b.   The following [18] annulene gives ^{1}H NMR signals at δ 3.0 and δ 9.3.

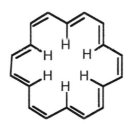

c.   The conjugated acid of the compound shown next is a weak acid (pK_a ~ 5–6). What is the structure of that weak acid? Compare the π systems of the two possible acids to explain your choice of structure.

2.   (continued)

d.   Conductivity measurements show that the dibromide that follows ionizes to give a dication. Sketch the ^{1}H NMR of the starting dibromide and the dication. Predict approximate chemical shifts and explain your predictions.

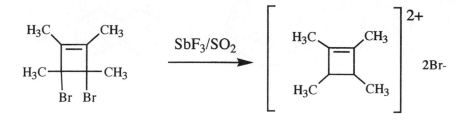

e.   Cyclopropenium ion is quite stable, but cyclopropene is not. Diphenylcyclopropenone has a dipole moment equal to 5.08 D, whereas benzophenone has a dipole moment equal to 2.97 D.

cyclopropenium   cyclopropene   cyclopropenone   benzophenone
      ion

2.   (continued)

    f.   The synthetic procedure shown here was a total failure; the starting material was recovered unreacted.

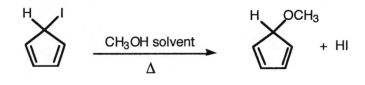

    g.   1,3-Cyclopentadiene is surprisingly acidic for a hydrocarbon (pK_a 15), whereas 1,3,5-cycloheptatriene is a very weak acid (pK_a 39).

3.   Compound **A**, when treated with aqueous sulfuric acid, was converted to isomeric
      compound **B** (C_9H_{12}). When **B** was heated for an extended time with chromic acid,
      dicarboxylic acid **C** was obtained. Give the structure of **B** and a mechanism that
      accounts for the formation of **B** from **A**.

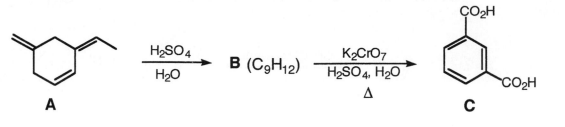

Reflection:

1.   Work with the other members of your group to construct a map that relates the concepts of aromaticity, antiaromaticity and nonaromaticity. The subsidiary concepts should be the structural, chemical, and physical properties that characterize the three concepts.

2.   The molecular orbital structure for cyclic, planar, fully conjugated systems consists of a single orbital with no nodes and pairs of isoenergetic orbitals with one node, with two nodes, etc.

      Give that molecular orbital structure, explain the origin of Hückel's rule. Apply these ideas to explain the experimental differences between cyclopentadiene anion and cation.

WORKSHOP 18

Aromatic Substitution

Purpose: The special properties of the aromatic ring and the patterns of reactivity of electrophiles with substituted benzenes were major challenges to the development of a mechanistic theory of organic chemistry. Key ideas about resonance and inductive effects were worked out and tested against the empirical patterns of substitution. The same ideas are the basis for understanding nucleophilic aromatic substitution. This Workshop will explore the mechanistic understanding that is the basis for the rational, nonempirical synthesis of polysubstituted aromatic compounds.

Expectations: You should be prepared to discuss and use the following ideas: resonance; inductive effects; electrophilic substitution; nucleophilic substitution; electrophilic addition; electron donating; electron withdrawing; Hammond principle; *o-*, *m-*, and *p-*; and regiochemistry.

1. Each of the reactions that follow is classified as an aromatic substitution. Construct a table identifying the electrophile, the nucleophile, and the intermediate for each reaction. Write generalized two-step mechanistic schemes (use curved arrows to show movement of electron pairs) for these reactions.

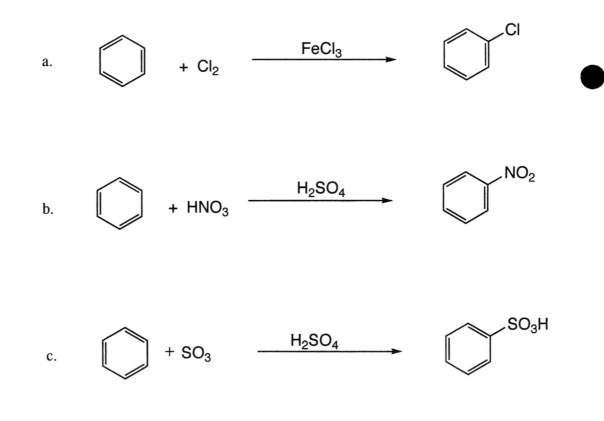

1.   (continued)

d.

e.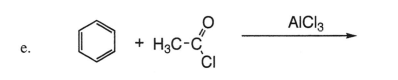

f.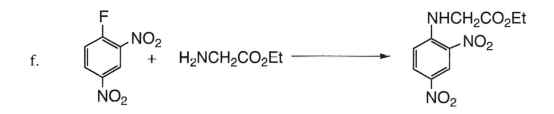

g.

2. Consider and explain the following differences in rate of reaction at 25°. Work in pairwise teams and make pairwise analyses: a *vs.* b; b *vs.* c; d *vs.* e; and f *vs.* g.

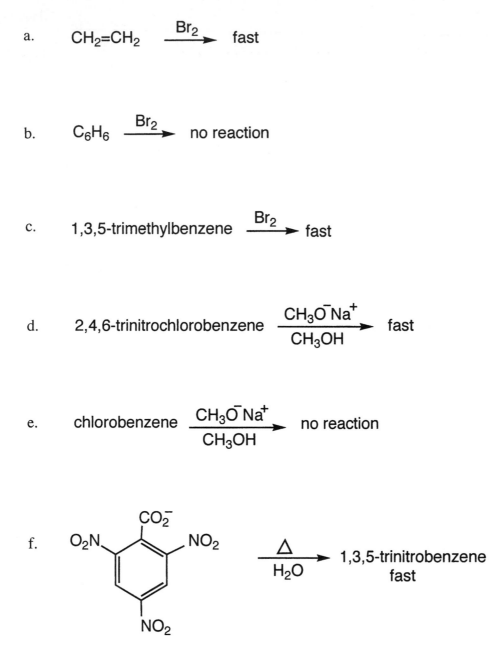

a. $CH_2=CH_2$ $\xrightarrow{Br_2}$ fast

b. C_6H_6 $\xrightarrow{Br_2}$ no reaction

c. 1,3,5-trimethylbenzene $\xrightarrow{Br_2}$ fast

d. 2,4,6-trinitrochlorobenzene $\xrightarrow[CH_3OH]{CH_3\bar{O}\,Na^+}$ fast

e. chlorobenzene $\xrightarrow[CH_3OH]{CH_3\bar{O}\,Na^+}$ no reaction

f. $\xrightarrow[H_2O]{\Delta}$ 1,3,5-trinitrobenzene fast

g. $\xrightarrow[H_2O]{\Delta}$ no reaction

3. a. Give multistep mechanisms for the reactions that follow. Be sure to distinguish the rate-determining and product-determining steps.

$(CH_3)_2C=CH_2 + DBr \longrightarrow (CH_3)_2\underset{\underset{Br}{|}}{C}CH_2D$

b. In each case, explain how the mechanism rationalizes the observed regiochemistry of the reactions.

3.   (continued)

c.   Give reaction-energy diagrams and verbal explanations that rationalize the differences in the product-determining steps of the two reactions.

4.   Consider the following experimental observations about the reaction of anisole with HNO_3/H_2SO_4:

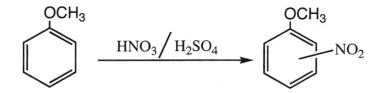

■   Anisole is several hundred times more reactive than benzene.

●   The composition of the product is approximately 70% *p*-, 29% *o*-, and 1% *m*-nitroanisole.

Use the Hammond principle to construct a reaction-energy diagram for the conversion of benzene or anisole to the intermediate cyclohexadienyl cations. Be sure to show the relative activation energies for the reaction of benzene and the reaction of anisole to give three different cyclohexadienyl cation intermediates.

5.   Consider the following observations:

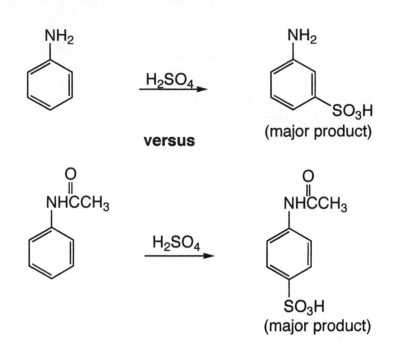

a.   Which reaction is faster? Explain your choice.

b.   Explain, with words and structures, the origins of the observed regiochemistry of the sulfonation reactions.

c.   Construct reaction-energy diagrams for these two reactions. In each case, compare the observed reactions with the reaction of benzene with H_2SO_4 under the same conditions. Be sure to show the relative activation energies for the formation of o-, p-, and m-sulfonation.

6.   Propose reagents and conditions for preparing the following compounds from benzene:

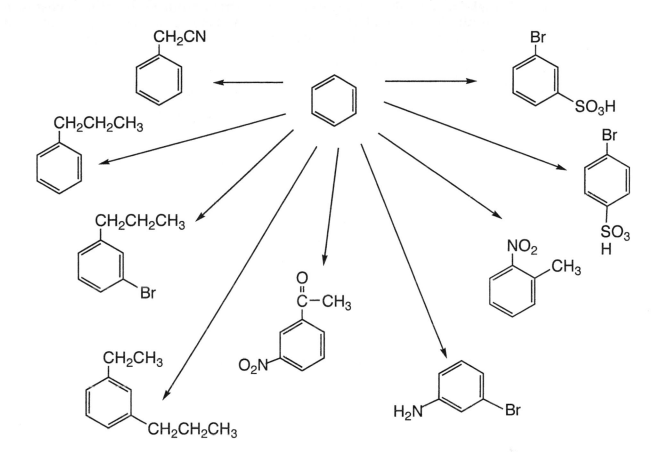

Reflection:

a.   Make a chart that segregates *o*-, *p*-directing from *m*-directing substituents. Identify the nonbonding electron pairs that increase the rate of *o*-, *p*-substitution or the partial positive charges that decrease the rate of *o*-, *p*- substitution.

b   Work with your colleagues to construct a flowchart that describes the sequential steps you follow to efficiently solve a synthesis problem in electrophilic aromatic substitution.

WORKSHOP 19

Pericyclic Reactions

Purpose: Much of our attention has been focused on polar reactions and radical reactions involving multistep processes and reactive ion or radical intermediates. In contrast, we have encountered a few reactions that proceed by way of one-step, elementary processes in which bond breaking and bond making occur in concert (i.e., simultaneously). Familiar examples of concerted reactions include S_N2 and E2 reactions and the additions of ozone, OsO_4 and MnO_4^- to double bonds. While S_N2 and E2 reactions fit easily into our understanding of polar reactions, the additions to double bonds to give cyclic products (cycloadditions) are more puzzling. In addition, you may have already encountered the Diels–Alder reaction, in which dienes and double bonds add to one another. The big surprise about these cycloadditions and other related reactions is that the concept of aromaticity and the 4n + 2 rule apply to transition states as well as to ground-state structures. You should emerge from this Workshop with a deeper understanding of concerted reactions and an insight into the puzzle about why some reactions occur and others do not. The Workshop provides another answer to the fundamental question, "What reacts with what?"

Expectations: You should review: OsO_4, MnO_4^-, and ozone additions; M.O.'s for butadiene and ethylene; Diels–Alder reactions and the stereochemistry of concerted additions; Cope and Claisen rearrangements.

1. Explain the observed reactivities and give structures for the expected products of the following reactions:

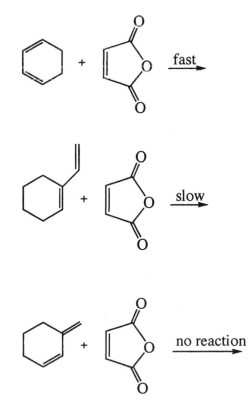

2.   a.   Diels–Alder reactions generally are interpreted as interactions between the HOMO of the diene and the LUMO of the dienophile. Predict the relative reactivities of the pairs of compounds that follow, and explain your reasoning. Give structures for the expected products, taking special care to describe the stereochemistry.

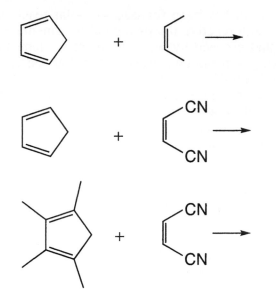

b.   It is possible to increase Diels–Alder reactivity with Lewis acid catalysis. Which orbitals are affected and why might this increase reactivity?

3. Construct M.O. diagrams for 1-3-butadiene and ethylene.

a. Use the symmetries (phases) of the interacting orbitals to show that the transition state for the Diels Alder reaction involves bonding–bonding interactions between the HOMO and the LUMO.

b. In contrast, the thermal cycloaddition of two ethylenes does not occur. Use the symmetries (phases) of the interacting orbitals to show that the transition state for this reaction involves an antibonding interaction between the HOMO and the LUMO.

3.   (continued)

c.   Our curved-arrow formalism shows us how many electrons are involved in these cycloaddition reactions. Characterize the Diels–Alder and the ethylene dimerization transition states as aromatic or antiaromatic.

d.   Hydride and alkyl shifts are fast signature reactions of carbocations.

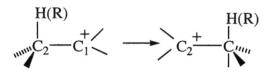

Give a representation of the interacting orbitals in the transition state for this reaction. How many electrons are involved in the cyclic transition state? Explain why this carbocation reaction is fast, but the corresponding carbanion rearrangement is not observed.

5. Ergosterol is a component of some of the vegetables in our diet. It is transformed by a photochemical reaction (under the skin) to precalciferol, a precursor to calciferol (vitamin D_2).

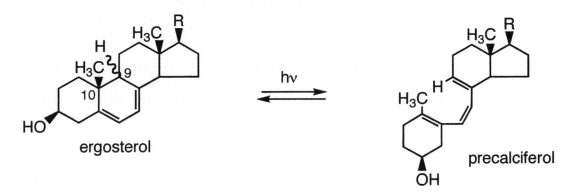

$$R = (Z)\text{-}CH(CH_3)CH=CHCH(CH_3)CH(CH_3)_2$$

Predict the relative stereochemistry of the –H at C_9 to the –CH$_3$ at C_{10} in ergosterol. Explain your reasoning.

6.   a.   1,5-Dienes undergo thermal rearrangement reactions as illustrated in the following diagram:

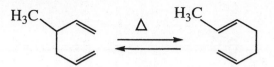

How many electrons are involved in the delocalized transition state?

b.   A particular example is the rearrangement of allyl vinyl ethers.   Give a representative structure and the expected product.

Predict the products expected from the following reactions:

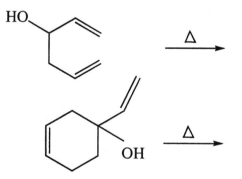

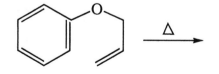

Reflection: Consider the observed stereochemistry of Diels–Alder reactions:

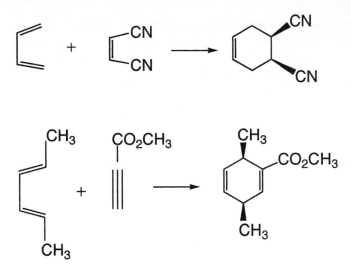

a.   Suppose that the Diels–Alder reaction involved a biradical intermediate:

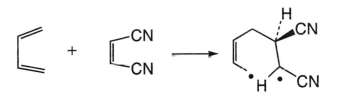

Explain how the observed stereochemistry of the D–A reaction makes this biradical mechanism unlikely.

Reflection: (continued)

b.   Construct reaction-coordinate diagrams to illustrate the differences between hypothetical stepwise and concerted mechanisms for a Diels–Alder reaction. Be sure to give specific structures for the energy maxima and minima on the diagrams. Specify the relative energies of the maxima for the rate-determining steps on the two reaction-coordinate diagrams.

c.   Otto Diels and his former student Kurt Alder shared the Nobel prize in 1950 for the reactions that bear their names. Brainstorm with your colleagues to identify reasons for according such special status to Diels–Alder reactions.

WORKSHOP 20

Aldehydes and Ketones

Purpose: Because of the polarization of the carbon–oxygen double bond, the chemistry of the carbonyl group is quite distinct from the chemistry of the carbon–carbon multiple bond. In general, alkenes (alkynes) function as a nucleophiles in reactions with electrophiles. In contrast, the polarized carbonyl group is the electrophilic partner in reaction with nucleophiles. The result is a wealth of new chemistry. You have already encountered some of this chemistry in the Workshop on alcohols. But there is more to be explored in this Workhop on nucleophilic additions to aldehydes and ketones. The formation of the tetrahedral intermediate and its subsequent reactions is a powerful mechanistic idea that unifies much of the chemistry of aldehydes, ketones, and the carboxylic acid derivatives, esters, amides, acyl halides, and anhydrides. Although the chemistry looks new, you should be able to integrate it with some aspects of your previous studies of nucleophilic substitution reactions.

Expectations: You should review the following concepts: polar bonds, HOMO and LUMO, π^* molecular orbital, amphoteric, nucleophile, basicity, catalysis, tetrahedral intermediate, and oxonium ion.

1. Consider the following experimental observations:

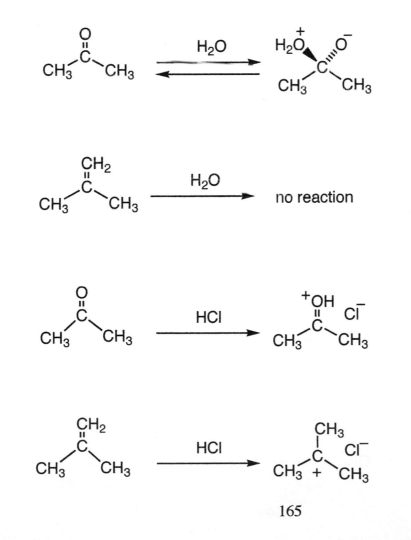

165

1.   (continued)

Compare and contrast the structures and reactivities of the double bonds of 2-methylpropene and 2-propanone, as illustrated in the preceding reaction diagrams. In each case, use the curved-arrow formalism to describe the bond-making and bond-breaking processes. Identify the important HOMO–LUMO interactions.

2.   For each pair of reactions that follows, circle the reaction that is *faster*. Explain your choices.

a. $(CH_3)_2C=O$ + H_2O

VS

$(CH_3)_2\overset{+}{C}=OH$ + H_2O $\longrightarrow$

b. $(CH_3)_2C=O$ + H_3N $\longrightarrow$

VS

$CH_3CH=O$ + H_3N $\longrightarrow$

c. $(CH_3)_2C=O$ + H_2O $\longrightarrow$

VS

$(CH_3)_2C=O$ + ^-OH $\longrightarrow$

3. Given what you already know about nucleophiles and leaving groups from your studies of S_N2 and $S_N1/E1$ reactions, make qualitative predictions about the reversibility of the reactions that follow. Characterize the ratio k_f/k_b of rate constants for the forward and back reactions as $>>1$, ~ 1, or $<<1$. Explain your choices.

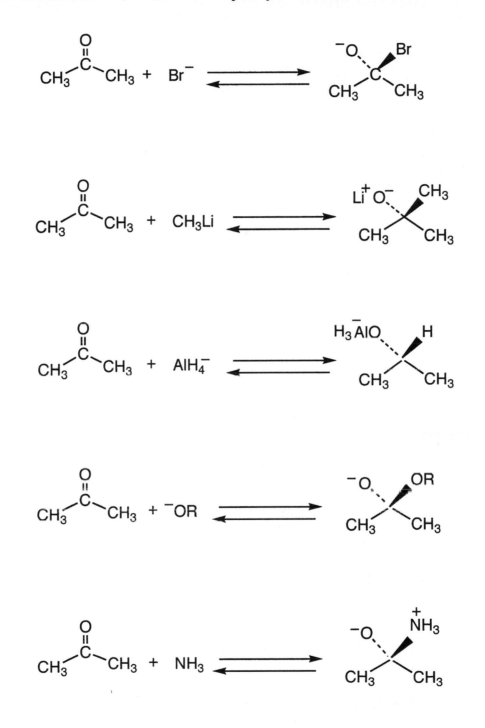

4.   Write multistep mechanisms to explain the experimental observations given next. Use the curved-arrow formalism to describe the bond making and bond breaking.

    a.   The labeling of acetone with ^{18}O is catalyzed by both acid and base. Be sure to explain the catalysis.

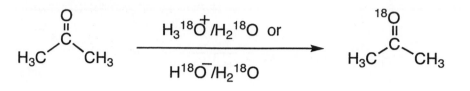

b.

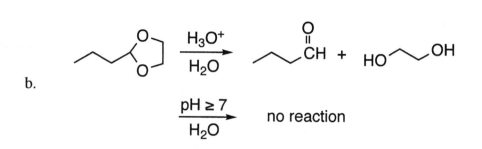

4.   (continued)

c.   The following reaction occurs rapidly at pH = 5, but fails at pH = 1 or pH = 8:

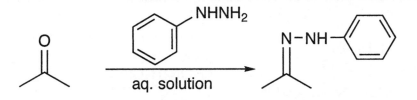

5.   Propose structures for the lettered compounds: Compound **A**, $C_9H_{12}O$, was optically active, did not react with 2,4-dinitrophenylhydrazine, showed a broad IR band at 3400 cm^{-1}, and was readily oxidized to **B**, $C_9H_{10}O$, with aqueous chromic acid (H_2CrO_4) at room temperature. When **A** was refluxed with chromic acid, benzoic acid was obtained. Compound **B** showed strong IR absorption at 1670 cm^{-1}, but none at 3400 cm^{-1}, and reacted with 2,4-dinitrophenylhydrazine. When **B** was reacted with EtMgBr followed by aqueous workup, compound **C** was obtained. **C** did not react with chromic acid at room temperature and showed an IR band at 3400 cm^{-1}. **B** and **C** were optically inactive and could not be resolved.

6. Glucose is a polyhydroxylic aldehyde. In aqueous solution, glucose is >99% in the cyclic hemiacetal form (no aldehyde C-H in the NMR spectrum at ~10ppm). Nevertheless, an aqueous solution of glucose reacts with hydroxylamine hydrochloride to give >90% yield of the corresponding oxime:

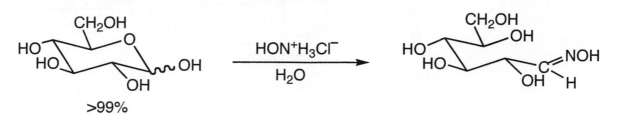

>99%

a. Write a mechanism for the formation of the oxime.

b. Explain why hydroxylamine hydrochloride, rather than hydroxylamine, is the reagent of choice.

c. Explain how the oxime formed in >90% yield when glucose is >99% in the cyclic, hemiacetal form.

7. Show how to carry out the specified chemical conversions, using any necessary organic and inorganic reagents. More than one step may be required.

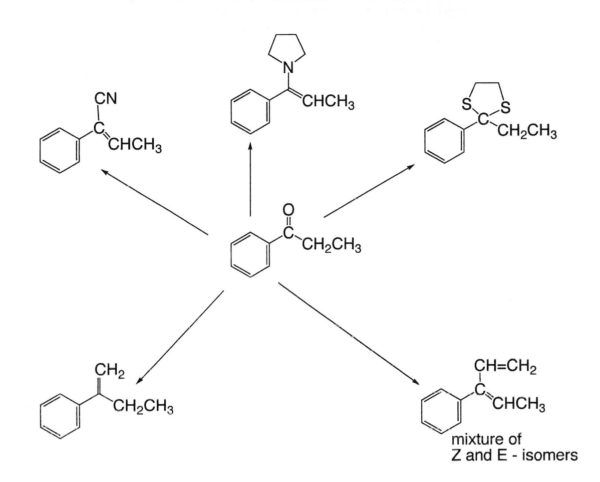

mixture of
Z and E - isomers

Reflection:

a. The principle of microscopic reversibility tells us that the transition state (i.e., the path) for the forward and reverse reactions must be identical. With this principle in mind, write a generalized mechanism for the acid-catalyzed hydrolysis of an enamine to its components: a secondary amine and a ketone.

b. Write a short paragraph explaining how hemiacetal and acetal formation and hydrolysis are prototypical reactions that provide the basis for understanding a host of other examples of additions to carbonyle (e.g., cyanohydrins, oximes and hydrazones, enamines, and the Wittig reaction).

Enols and Enolate Ions

Purpose: The carbonyl functional group is the all-time favorite of synthetic chemists. We saw part of the reason for that in the preceding Workshop: A wide variety of nucleophiles add to the carbonyl group, allowing the construction of new C-O, C-N, C-S, and C-H bonds. But there is even more: Carbanions are easily formed at the carbons immediately adjacent to the carbonyl group. In turn, these carbonions are excellent nucleophiles for new bond-forming reactions at the α-carbons. We had a preview of these carbanions in our consideration of conjugated systems and in the enol ⇌ ketone interconversion that is involved in the hydration of alkynes. In this Workshop, we will expand the utility of the carbonyl group to include the reactions of the corresponding carbanions (enolate ions) and enols.

Expectations: Review carbonion structures, pK_a, enol–keto equilibrium, allylic systems, molecular orbitals for 3-adjacent p-orbitals, resonance structures, racemization, kinetic vs. thermodynamic control, aldol reactions, and retrosynthesis.

1. Consider the following pK_a values, and provide explanations (verbally and with structures) for the breaks labeled A, B, and C:

CH_3CH_2—H	50
	_____ A
$CH_2=CHCH_2$—H	43
$\bigcirc$— CH_2—H	41
	_____ B
$C_2H_5OCCH_2$—H ($\overset{\parallel}{O}$)	24
CH_3CCH_2—H ($\overset{\parallel}{O}$)	19.3
$HCCH_2$—H ($\overset{\parallel}{O}$)	16.7
O_2NCH_2—H	10.2
	_____ C
$CH_2=CHO$—H	10.5
$\bigcirc$—O—H	10.0

2.   Three isomeric compounds, $C_{12}H_{16}O$—say, **A**, **B** and **C**—were all optically active, and all had strong IR bands in the IR 1700 cm $^{-1}$ region. When **A** was treated with NaOD/D_2O, three hydrogens were exchanged for deuterium (D), and the deuterated A remained optically active. Compound **B** exchanged only one hydrogen when treated with NaOD/D_2O, and the monodeuterated **B** was then optically inactive. When Compound **C** reacted with NaOD/D_2O, four hydrogens were exchanged for D's, and the deuterated **C** remained optically active. **A**, **B**, and **C** all had multiplet absorptions at 7–8 ppm (5H) in the NMR. Give structures for **A–C** consistent with the preceding information. (There may be more than one structure possible for **A**, **B**, or **C**.)

3.   Give mechanisms for each of the following reactions, clearly showing all important intermediates and using curved arrows to show movement of electron pairs:

a.   (R)-2-methylcyclohexanone $\xrightarrow[\text{in } H_2O]{^-OH \text{ cat.}}$ (R,S)-2-methylcyclohexanone

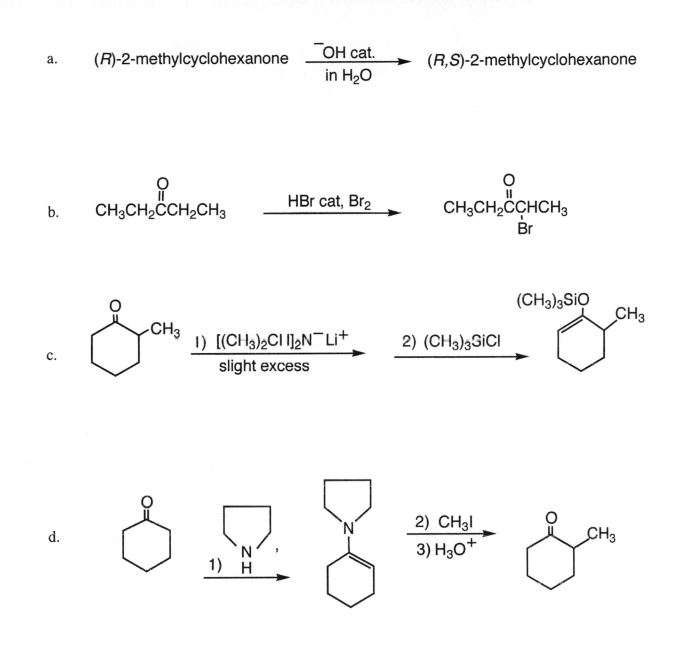

b.
$$CH_3CH_2\overset{\overset{\displaystyle O}{||}}{C}CH_2CH_3 \xrightarrow{\text{HBr cat, } Br_2} CH_3CH_2\overset{\overset{\displaystyle O}{||}}{C}\underset{\underset{\displaystyle Br}{|}}{CH}CH_3$$

c.   1) $[(CH_3)_2CHI]_2N^- Li^+$ slight excess   2) $(CH_3)_3SiCl$

d.   1) $\overset{}{\underset{H}{N}}$,   2) CH_3I 3) H_3O^+

4. The alkylation of unsymmetrical ketones such as 2-methylcylohexanone often gives mixtures of products, including 2,2-dimethylcyclohexanone, 2,6-dimethylcyclohexanone, and even tri- and tetramethylcyclohexanone:

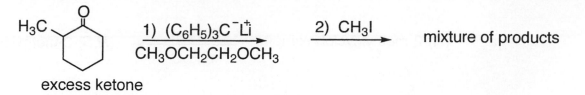

excess ketone

As a result, synthetic chemists have developed techniques for specific alkylations. For example, when the foregoing reaction is repeated with the base in excess, 2,6-dimethylcyclohexanone is formed in high yield:

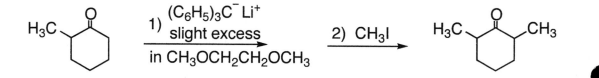

Explain why the presence of excess ketone leads to nonspecific alkylation and a mixture of products. (*Hint*: $(C_6H_5)_3C^-$ Li^+ is a very strong base, so the reaction with the ketone is irreversible. Nevertheless, are there other possible acid–base reactions in the presence of excess ketone?)

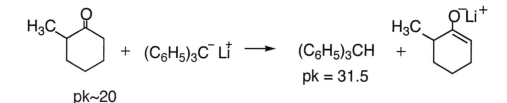

5. a. Many additions to the carbonyl group, including acetal formation and the aldol reaction, are reversible. For most ketones, the equilibrium constants favor the starting ketone. Explain how the reaction conditions can be manipulated to give useful yields of acetal and aldol products.

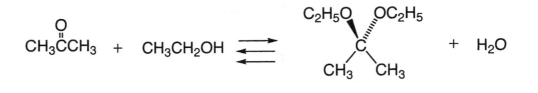

b. Propose a mechanism for the following reaction:

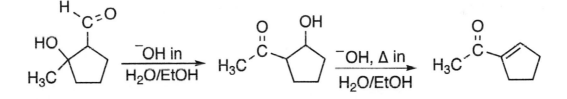

6. Disconnect the following molecules to make simpler precursors suitable for C-C bond-forming reactions:

a.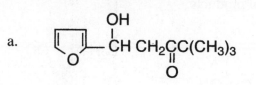

OH

b. $C_6H_5CH = C \overset{NO_2}{\underset{CH_2CH_3}{\diagup}}$

c.

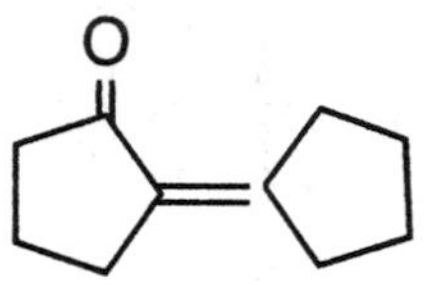

d. 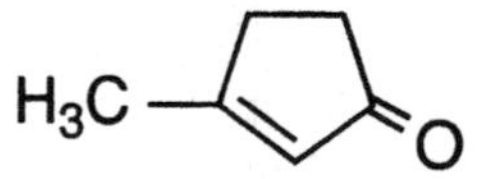 H_3C —

e. 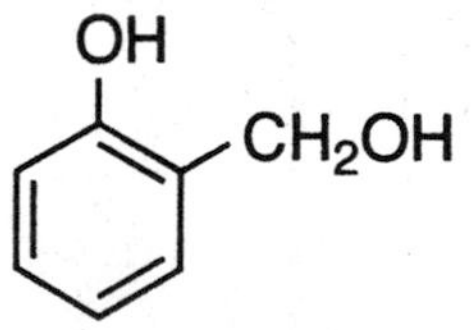 OH

CH$_2$OH

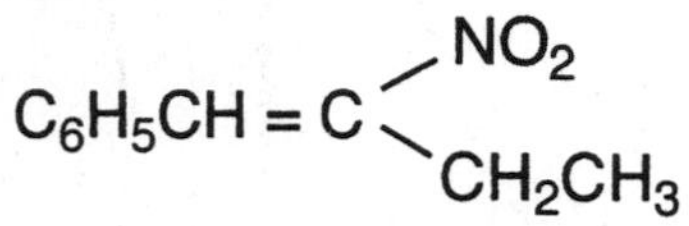

7.   Show how to carry out the following chemical conversions, using any necessary organic
       and inorganic reagents (more than one step may be required):

a.

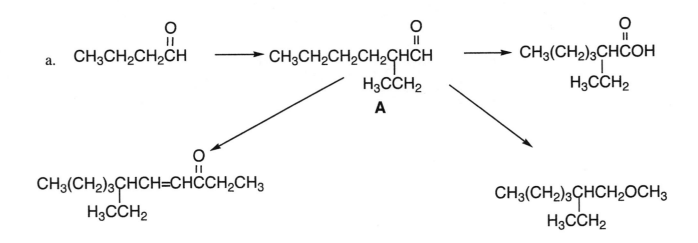

b.

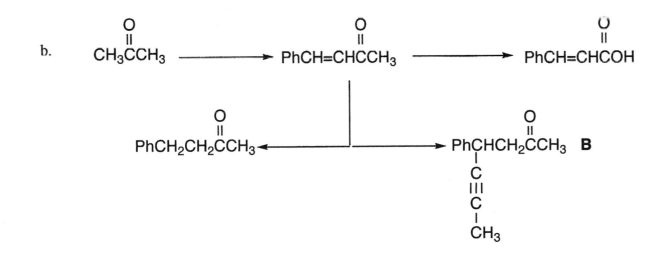

Reflection: For each of the reactions in Problem 6, and for the structures labeled **A** and **B** in Problem 7, identify the nucleophile and the electrophile that react in the C-C bond-forming step. Then organize a retrosynthesis map that shows how various structures can be built.

e.g.,

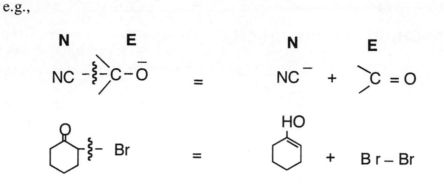

WORKSHOP 22

Ester and β-Dicarbonyl Enolates

Purpose: Here are some suggestive pKa's:

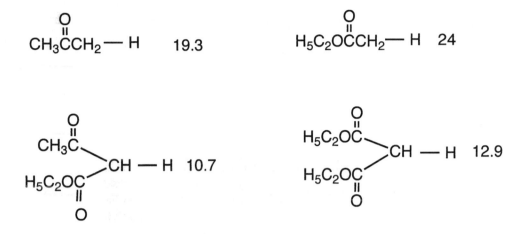

The pKa for ethyl acetate forecasts a chemistry for ester enolates and enols similar to the chemistry of aldehydes and ketones. Because the ester is at an oxidation state different from that of the aldehyde or ketone, the products of the ester enolates are also at a different oxidation state. The enolate ions from β-dicarbonyl compounds are even more accessible than enolate ions from aldehydes, ketones, and esters. The ester function of β-keto esters is roughly the inverse of a protecting group: It serves to activate the adjacent methylene and is then discarded after it has done its job. The enolate ion is undoubtedly the most important reactive intermediate in synthetic organic chemistry, and this Workshop continues our exploration of the formation and reactions of these nucleophiles.

Expectations: You should be familiar with the following terms and concepts: Claisen reaction, acetoacetic ester, diethyl malonate, decarboxylation, cyclizations, Michael reaction, enamines, and Robinson annulation. Be prepared to use your book to look up puzzling reactions.

181

1. Provide structures for the major products of each of the reactions that follow. In each case, identify the electrophile and the nucleophile in the carbon–carbon bond-forming steps. Show the flow of electrons, using the curved-arrow formalism.

a.

b. PhCH + CH₃CCH₂CO₂Et —Et₂NH / EtOH→

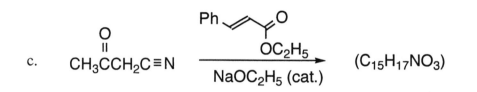

c. CH₃CCH₂C≡N —— / NaOC₂H₅ (cat.)→ (C₁₅H₁₇NO₃)

d. CH₃CH₂CO₂Et   1. [(CH₃)₂CH]₂N⁻ Li⁺ / 2. CH₃CH₂I

2.   a.   Give a reasonable mechanism for the reaction that follows, clearly showing the structures of all important intermediates. Use curved arrows to show the bond making and bond breaking.

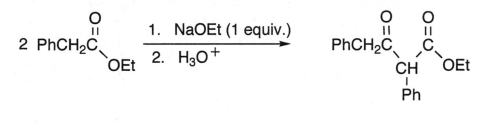

b.   Consider the Claisen reaction. The C-C bond-forming step is reversible, and the equilibrium constant favors the reactants. Explain clearly how the structure of the ester and the concentration of the base are controlled to drive the reaction to completion. What would happen if a catalytic amount of NaOEt were used? What would happen if the starting ester were Ph_2CHCO_2Et?

2.   (continued)

    c.   When compound **C** in ethanol is stirred with one molar equivalent of $NaOC_2H_5$ and one equivalent of $CH_3CH_2CH_2Br$, a new compound, **D**, is formed. When **D** is refluxed with 1.5-**M** aqueous H_2SO_4, the ketone **E** is formed. Propose a structure for **D**, and give a stepwise electron-pushing mechanism leading from **C** to **D** to **E**.

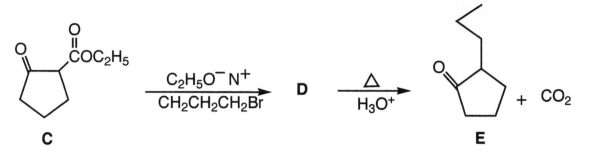

3.   Disconnect each of the compounds that follow into components that would assemble to give the designated compounds. Use the electron-pushing formalism to show the bond-forming steps.

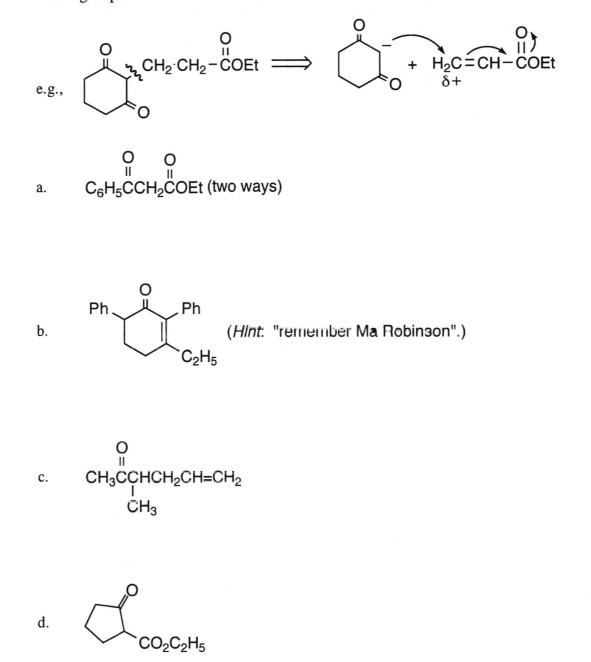

e.g.,

a. $\overset{O}{\underset{||}{C}}_6H_5\overset{O}{\underset{||}{C}}CH_2\overset{O}{\underset{||}{C}}OEt$ (two ways)

b.   (*Hint*: "remember Ma Robinson".)

c. $CH_3\overset{O}{\underset{||}{C}}\underset{\underset{CH_3}{|}}{C}HCH_2CH{=}CH_2$

d.

4.   Having available ethyl acetoacetate, diethyl malonate, benzene, any compounds with three or fewer carbons, and any inorganic reagents, show how to synthesize the following compounds:

a.
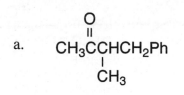

b.
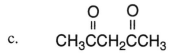

c.
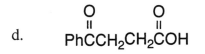

d.
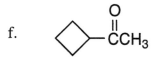

e. $CH_3CCH_2CH_2CH_2CH_2CH_2CCH_3$
         $\underset{O}{\|}$ $\underset{O}{\|}$

f.
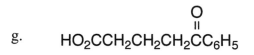

g. $HO_2CCH_2CH_2CH_2\overset{\displaystyle O}{\overset{\|}{C}}C_6H_5$

Reflection:

a.   As in the previous Workshop, identify the electrophile and the nucleophile in the various reactions in this Workshop and organize a retrosynthesis map that shows how different structures can be built.For example, a β-diketone such as the one in problem 3a can be built in two different ways, corresponding to the reactions of two different ketones (the nucleophiles) and two different esters (the electrophiles). Work out these two retrosynthetic paths for β-diketones. Then generalize the retrosynthetic process to build other structures, such as methyl ketones, carboxylic acids, 1,4-dicarbonyl compounds, and β-keto esters and rings, as illustrated in Problems 3 and 4.

c.   Choose representative reactions to illustrate the introductory comment about different oxidation states at the beginning of this Workshop. In particular, explain how the oxidation state of the product can tell you which kind (oxidation state) of starting material is best.

WORKSHOP 23

Carbohydrates

Purpose: Carbohydrates are multifunctional polyhydroxy aldehydes or ketones. Like other multifunctional compounds, they show reactions that are characteristic of their individual functional groups. More interesting, however, are the intra- and intermolecular reactions of the two functional groups to give cyclic and polymeric products, respectively. This Workshop provides satisfying opportunities to apply ideas about the formation and reactions of hemiacetals and acetals to unravel much of the chemistry of the carbohydrates. In addition, the structures are complex, with multiple stereogenic centers. As a result, the Workshop will also give you a chance to revisit and apply ideas about conformational and configurational stereoisomers. Finally, it is especially satisfying to apply our fundamental ideas about organic chemistry to important bioorganic molecules. Carbohydrates are ubiquitous in living systems, serving a host of essential functions from building and maintaining structure units, to energy storage, to molecular recognition.

Expectations: Prior to beginning this Workshop, review the chemistry of alcohols, aldehydes, and ketones; hemiacetals and acetals; acid–base chemistry; aldol and retro-aldo chemistry; the structure and reactions of carbohydrates; and conformational analysis and stereochemistry (enantiomers and diastereomers). Bring your textbook to the Workshop to look up structures of simple sugars such as D-arabinose and D-mannose.

1. **Reminder:** Fischer projection is a convention for representing 3-D molecules on 2-D surfaces. As with all conventions, there are specific rules to be obeyed. (*R*)-Glyceraldehyde is the simplest monosaccharide. *Convert the 3-D representation into a Fischer projection formula.*

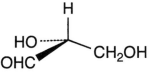

(*R*)-glyceraldehyde

D-glyceraldehyde

2. Provide structures that satisfy the following descriptions:

    a. The most stable chair conformation of α-D-mannopyranose. (*Hint*: Start from the observation that all of the substituents are equatorial in the most stable chair conformation of β-D-glucopyranose.)

188

2.   (continued)

b.   The enediol that is an intermediate in the base-catalyzed isomerization of the carbonyl group of D-arabinose from the 1-position to the 2-position.

3.   Provide structures for the major organic products of the following reactions:

a.

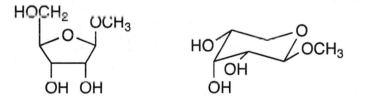

methyl-β-D-galactopyranoside

H_3O^+ / H_2O → (give Fischer projection)

b.   D-mannose $\xrightarrow[\text{CH}_3\text{CH}_2\text{OH}]{H_3O^+}$ (give chair conformation of product(s))

4.   When D-ribose is treated with methanol and HCl, a mixture of glycosides is formed. Two of these compounds are shown here. Describe both verbally and with chemical equations how periodic acid could be used to distinguish between the two compounds.

5. Optically active carbohydrate **A** ($C_{14}H_{20}O_7$) does not react with Cu(II) complex and, upon treatment with dilute HCl, gives optically active carbohydrate **B** ($C_7H_{14}O_7$) plus benzyl alcohol. Reaction of **A** with excess dimethyl sulfate gives **C**, which, upon dilute acid hydrolysis, yields the optically active 2,3,5,6,7-pentamethyl aldoheptose **D**. When **B** is treated with nitric acid, optically inactive **E** is formed. The oxidative removal of C_1 of **B** produces D-glucose. Give structures **A** through **E** consistent with this information.

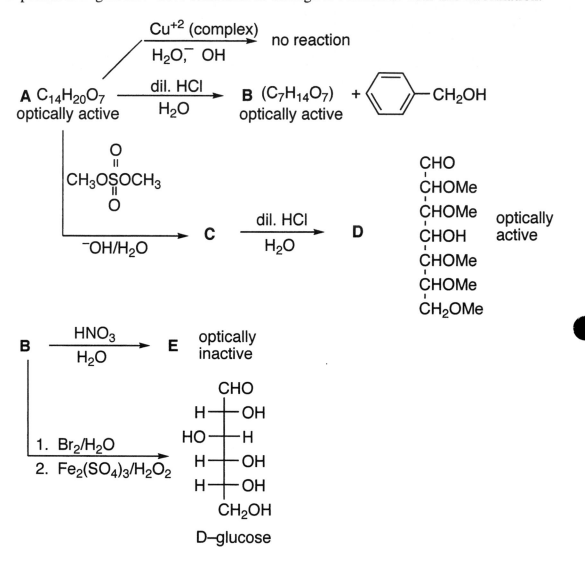

6. When D-fructose is treated with dilute aqueous hydroxide ion, D-glyceraldehyde and dihydroxyacetone are formed.

   a. Both verbally and using structures (mechanisms), clearly explain how the two trioses are generated from D-fructose.

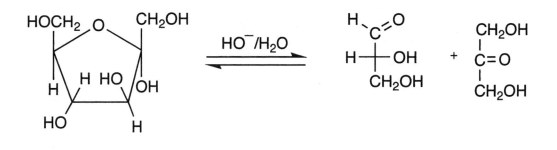

   b. In principle, the D-glyceraldehyde and dihydroxyacetone can undergo the reverse reaction under basic conditions to form not only D-fructose, but also three other ketohexoses that are diastereomeric with D-fructose. Give structures for these three ketohexoses. (Use a Fischer projection.)

7. Many complex oligosaccharides have key roles in molecular recognition. Oligosaccharides that are specific antigens on the surfaces of pathogens or tumor cells are especially interesting. (See Figure 1.) When synthetic antigens are injected into the human body, the immune system reacts to generate antibodies. Therefore, these synthetic oligosaccharides are potential vaccines against pathogens or even cancer. This is currently an area of intense research.

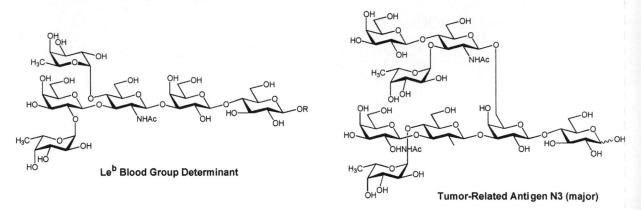

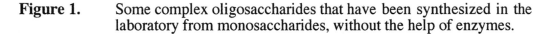

Figure 1.   Some complex oligosaccharides that have been synthesized in the laboratory from monosaccharides, without the help of enzymes.

An important method for synthesizing oligosaccharides from monosaccharides involves nucleophilic substitution to open an epoxide. As a model for this key step in this "glycal assembly" reaction, consider the acid-catalyzed reaction of methanol with the epoxide of dihydropyran (**1**) to give **3**:

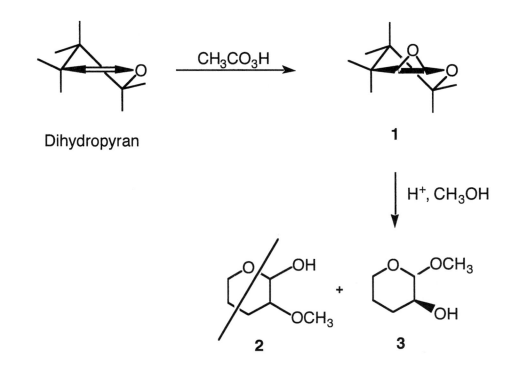

7.   (continued)

a.   Only **3** is formed. Account, in as clear a manner as possible, for this observed regiochemistry

b.   Give the chair structure of the first-formed stereoisomer of **3**.
        (Recall Workshop 15.)

c.   Give the most stable chair structure of **3**.

d.   The glycal assembly coupling step shown next can be used for a disaccharide synthesis or as an initial step in the synthesis of a large, complex oligosaccharide. The large ball on **4** represents either a protecting group or the attachment of the C6-hydroxyl to a support for solid-phase synthesis. A glycal is an unsaturated sugar, such as **5**. Show the structure of the first-formed disaccharide product from the reaction of **4** and **5**, clearly specifying both regio- and stereochemistry. (You can abbreviate **5** as ROH).

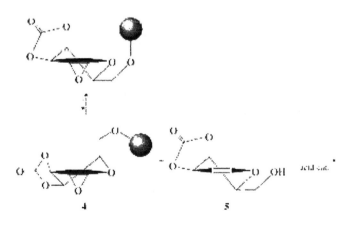

7.   (continued)

   e.   Show the structure of the most stable chair conformation of the disaccharide product.

   f.   Explain why a glycal is an appropriate nucleophile for the construction of a polysaccharide.

(Our thanks to Andreas Franz for this problem.)

Reflection: Carbohydrate chemistry often has its own language. Work with your colleagues to construct a table that links the special carbohydrate language to a simpler functional group language or reaction language.

Carbohydrate	is	a polyhydroxy aldehyde or ketone
D-sugar	is	(R)-configuration at the penultimate carbon
Aldoheptulose	is	.....
Epimer	is	.....
Reducing Sugar	is	.....
Etc.		.....

WORKSHOP 24

Phenols

Purpose: A phenol may look like an alcohol, but it makes a big difference whether the hydroxyl group is bonded to an aryl or an alkyl group. Phenols have distinctive properties and reactions of the hydroxyl group and the aromatic ring. This Workshop focuses on these properties.

Expectations: You should review the properties of the phenols in the context of electrophilic aromatic substitution, acidity, aromaticity, resonance, redox reactions, electrochemistry, the Diels–Alder reaction, free-radical reactions, benzyne, and nucleophilic additions.

1.   a.   Explain clearly the origins of the huge difference in the acidities of phenol ($pKa = 9.9$) and cyclohexanol ($pKa = 18$).

    b.   The oxidation of unsaturated fats, oils, and fatty acids is a free-radical reaction at the allylic C-H bond. The chain steps are as follows, where

$$ROO\bullet \ + \ RH \longrightarrow ROOH \ + \ R\bullet$$

$$R\bullet \ + \ O_2 \longrightarrow ROO\bullet$$

$$RH = -\underset{\underset{H}{|}}{C}\diagup^{\displaystyle CH=CH-}$$

In some cases, the hydroperoxides are useful intermediates (e.g., in the biosynthesis of prostaglandins); in other cases, they are sources of destructive •OH radicals. The oxidative chain reaction is terminated by phenols, but not by alcohols. Account for this difference as clearly as you can. (*Hint*: Predict, in a qualitative manner, the differences in BDE's for ArO−H and RO−H).

1.   (continued)

c.   Account for the different results in the following ether cleavage reactions:

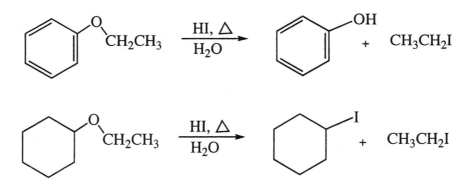

d.   Consider the differences in proton affinities of anisole and dimethyl ether, as revealed by their respective pKa's:

(CH$_3$)$_2$Ö$-$H

pKa = -3.8   pKa = -6.5

2. Give a reasonable mechanism for each of the reactions that follow. Use curved arrows to
    show movement of electron pairs. In each case, give an analogy to a simpler reaction that
    does not involve a phenol or a quinone.

a.
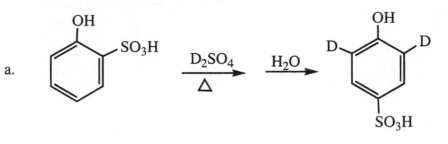

b.
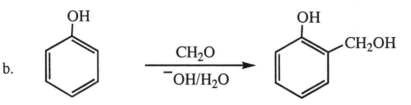

c.
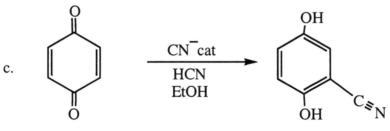

d.
 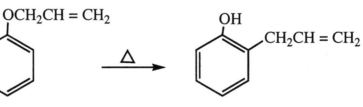 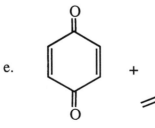

e.
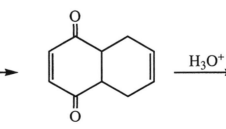 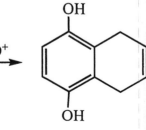

3. Phenol is an important industrial chemical that can be synthesized from benzene. One commercial method (Scheme 1) involves converting benzene to chlorobenzene and reacting chlorobenzene at high temperature and pressure with aqueous sodium hydroxide, followed by neutralization with acid.

Scheme 1

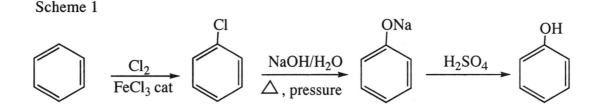

Another commercial method (Scheme 2) involves the acid-catalyzed reaction of benzene with propene (obtained in addition to the $C_5 - C_{10}$ gasoline fraction when the high-molecular-weight hydrocarbons from petroleum are catalytically cracked) to form isopropylbenzene. The reaction of isopropylbenzene with oxygen gives the corresponding hydroperoxide, which, when treated with aqueous strong acid, yields phenol plus acetone. (The commercial demand for acetone is comparable to that for phenol.)

Scheme 2

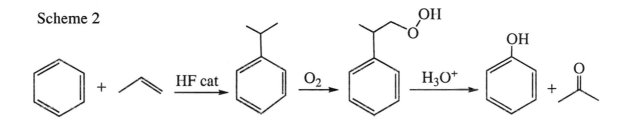

a. Balance each of the reactions in both Scheme 1 and Scheme 2, and give approximate reaction conditions.

3.   (continued)

   b,   Give the mechanism for each of the reactions in Scheme 1 and Scheme 2.

   d.   "Green chemistry" (environmental chemistry) emphasizes chemical reactions that are environmentally sound, synthetically efficient, energy efficient, and safe. Examine each of the preceding two industrial reaction schemes, and discuss how they measure up to the standards of green chemistry.

   e.   Choose which scheme represents the best green-chemistry synthesis of phenol, and justify your choice.

4.   Recall that one answer to the big question of what reacts with what is that *reducing agents react with oxidizing agents*. Consider the reaction of Ag⁺ with hydroquinone (QH₂) under standard conditions:

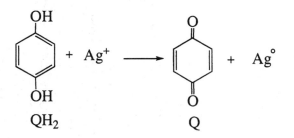

QH₂   Q

a.   Using the appropriate half-cell reactions, balance this redox reaction. Be sure to balance electrons, atoms, and charges.

b.   Some standard reduction potentials are as follows:

$$Q + 2H_3O^+ + 2e^- \rightleftharpoons QH_2 + 2H_2O \qquad \varepsilon° = 0.699V$$

$$Ag^+ + e^- \rightleftharpoons Ag° \qquad \varepsilon° = 0.799V$$

$$Sn^{4+} + 2e^- \rightleftharpoons Sn^{2+} \qquad \varepsilon° = 0.139V$$

$$Cu^{2+} + e^- \rightleftharpoons Cu^+ \qquad \varepsilon° = 0.161V$$

$$Cr_2O_7^{2-} + 6e^- + 14H_3O^+ \rightleftharpoons 2Cr^{3+} + 21H_2O \qquad \varepsilon° = 1.36V$$

Recalling that the more positive half-cell potential corresponds to the cathode (reduction) in the electrochemical cell, predict qualitatively whether Q or QH₂ will be favored at equilibrium with Ag⁺, Sn⁴⁺, Cu²⁺, and Cr₂O₇²⁻ under standard conditions. Stated differently, which of these metal ions would you choose to prepare quinone from hydroquinone? to prepare hydroquinone from quinone? Explain your reasoning.

Reflection: Use this Workshop to review fundamental ideas presented earlier about reactivity, structure, and mechanisms. Work with your colleagues to analyze each problem to find the connections to previous ideas.

WORKSHOP 25

Carboxylic Acids

Purpose: The carboxylic acid structure is amphoteric, and the acid–base properties are the most obvious and important reactions of this functional group. Many of our fundamental ideas about the effects of substituents on rates and equilibria are grounded in the effects of substituents on the acidities of carboxylic acids. This Workshop will help you understand the origins of those ideas.

Nucleophilic additions to the carbon–oxygen double bond of carboxylic acids are generally unproductive, because hydroxide ion is a poor leaving group. As a result, the most successful additions to the carbon–oxygen double bond are acid catalyzed. The change in the leaving group from hydroxide to water was previously observed in the acid-catalyzed reactions of alcohols. As in the chemistry of alcohols, several other reactions of carboxylic acids convert the –OH into good leaving groups. As a result, nucleophilic addition to the carboxylic acid group is rendered productive and leads to the preparation of acyl derivatives with the general structure RCX. This

$$\overset{\|}{O}$$

Workshop provides a good opportunity to make integrating connections among the chemistry of alcohols, carboxyl groups, and carboxylic acids.

Expectations: You should review observations and ideas about pKa; solubility; inductive effects; leaving groups in S_N2, E2, and S_N1/E1 reactions; the tetrahedral intermediate in nucleophilic additions to carbonyl groups; acetal formation and hydrolysis; oxidation–reduction; ozonolysis; and Grignard reactions.

1. For each of the pairs that follow, specify the member that is more acidic. Explain your choice verbally and with the help of structures.

    a. CH_3COOH versus FCH_2COOH

    b. $CH_2{=}CHCH_2COOH$ versus $HC{\equiv}CCH_2COOH$

    c. $CH_3CH_2CO_2H$ versus $HO_2CCH_2CO_2H$ (1st and 2nd ionization)

    d. $N{\equiv}CCH_2COOH$ versus $HC{\equiv}CCH_2COOH$

1.   (continued)

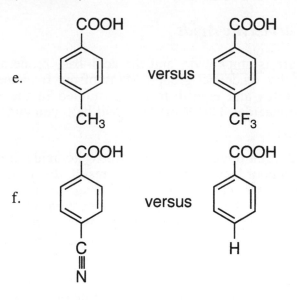

e.   COOH / CH₃   versus   COOH / CF₃

f.   COOH / C≡N   versus   COOH / H

2.   Predict the relative rates of nitration for the substituted arenes related to Problems 1(e) and 1(f); that is, which member of the following pairs reacts faster? Explain your choices.

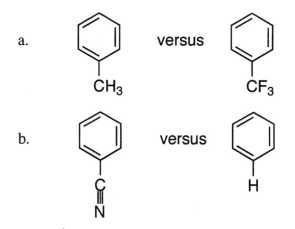

a.   CH₃   versus   CF₃

b.   C≡N   versus   H

How do the effects of a substituent on the acidities of benzoic acids compare with the rates of electrophilic substitution? Formulate a qualitative, predictive rule of thumb. Explain the structural basis for your rule.

3. Predict the major species at equilibrium. Justify your choices.

a. $C_6H_5CO_2H$ + H_2O $\rightleftharpoons$ $C_6H_5CO_2^-$ + H_3O^+

b. $CH_3CO_2^-$ + $C_2H_5O_2CCH_2CO_2C_2H_5$ $\rightleftharpoons$ CH_3CO_2H + $C_2H_5O_2C\overset{-}{C}HCO_2C_2H_5$

c. $CH_3CH=C\overset{\displaystyle OH}{\underset{\displaystyle OH}{<}}$ $\rightleftharpoons$ $CH_3CH_2\overset{\displaystyle O}{\overset{\|}{C}}-OH$

4. Compound **A**, $C_{12}H_{16}O_2$, exhibited a strong absorption in the IR spectrum at 1715 cm^{-1}. Treatment of **A** with NaOH/H_2O gave a neutral compound **B** and a sodium salt that, after acidification, gave compound **C**. **B** exhibited strong broad absorption in the IR spectrum at 3333 cm^{-1} and could be oxidized with aqueous chromic acid (Na$_2$Cr$_2$O$_7$/H$_2$SO$_4$/H$_2$O) to 2-methylbutanoic acid. **C** exhibited absorption in the IR spectrum at 1681 cm^{-1} and a very broad band over the range 2500–3500 cm^{-1}. The ^{1}H NMR spectrum of **C** exhibited absorption at δ 7.1–8.5 (m, 5H) and 12.70 (s 1H). Provide structures for **A**, **B**, and **C**.

5.   Construct a flow diagram showing how products of hydrolysis of each of the following
      esters with aqueous acid can be separated by acid–base extraction techniques:

i.

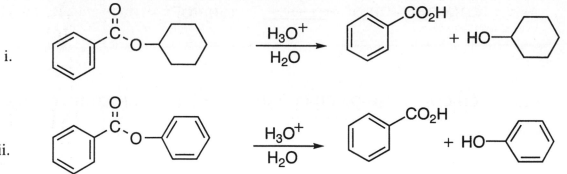

ii.

5.   (continued)

Give detailed mechanisms for the reactions that follow. Be careful to keep track of proton transfers, and use the curved-arrow formalism to show bond-making and bond-breaking processes.

a. $RCO_2H + H_2{}^{18}O \; \underset{\longleftarrow}{\overset{H_3O^+}{\longrightarrow}} \; R-C{\overset{{}^{18}O}{\underset{{}^{18}OH}{\big\langle}}} + H_2O$

b. $RCO_2H + R'OH \; \underset{\longleftarrow}{\overset{H_3O^+ \; cat}{\longrightarrow}} \; R\overset{O}{\overset{\|}{C}}OR' + H_2O$

c. $3RCO_2H \; \xrightarrow{PBr_3} \; R\overset{O}{\overset{\|}{C}}Br + P(OH)_3$

6.   Show how to carry out the synthetic conversions that follow. Specify necessary reagents and conditions.

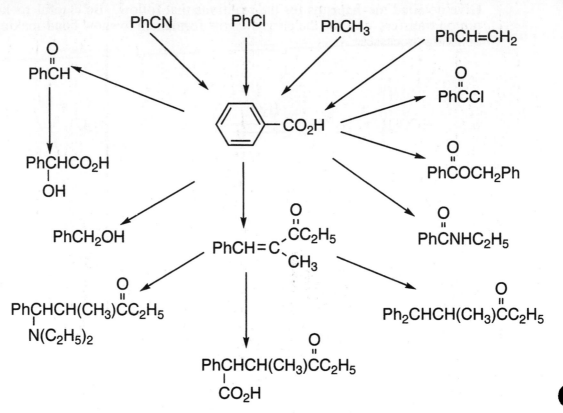

Reflection: Work with your colleagues to construct a reaction map for carboxylic acids that identifies the various chemical tactics for changing the hydroxyl group to a better leaving group.

WORKSHOP 26

Carboxylic Acid Derivatives: Nucleophilic Acyl Substitution

Purpose: At first glance, the structures and reactions of carboxylic acid derivatives seem bewildering. On closer inspection, however, a unifying mechanistic scheme emerges. Even better, the general mechanism turns out to be a variation on the mechanism of nucleophilic additions to aldehydes and ketones: Differences in structure, reactivity, and reaction pathway are simply consequences of differences in leaving groups. This Workshop provides opportunities to explore the variations in structure and reactivity of carboxylic acid derivatives, to build an integrated mechanistic understanding, and to put that insight to use.

Expectations: You should review ideas about leaving groups in nucleophilic substitution and elimination reactions (S_N2, E2, S_N1/E1), IR absorption for carbonyl groups, nucleophilic addition reactions and mechanisms, the "tetrahedral intermediate," and acid and base catalysis.

1.   Each of the reactions that follow involves nucleophilic substitution at an acyl carbon. For each reaction, construct a table showing the electrophile and the nucleophile in the key bond-forming step, the corresponding tetrahedral addition intermediate, and the leaving group in the bond-breaking step. Write a general mechanistic scheme (use curved arrows to show bond-making and bond-breaking processes) for these reactions, keeping in mind that acid- and base-catalyzed processes will differ in their timing of proton transfers.

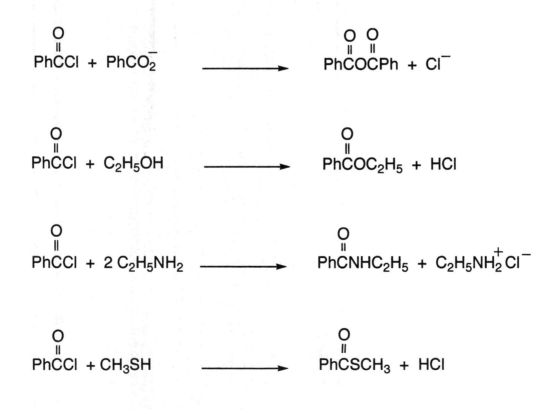

1.   (continued)

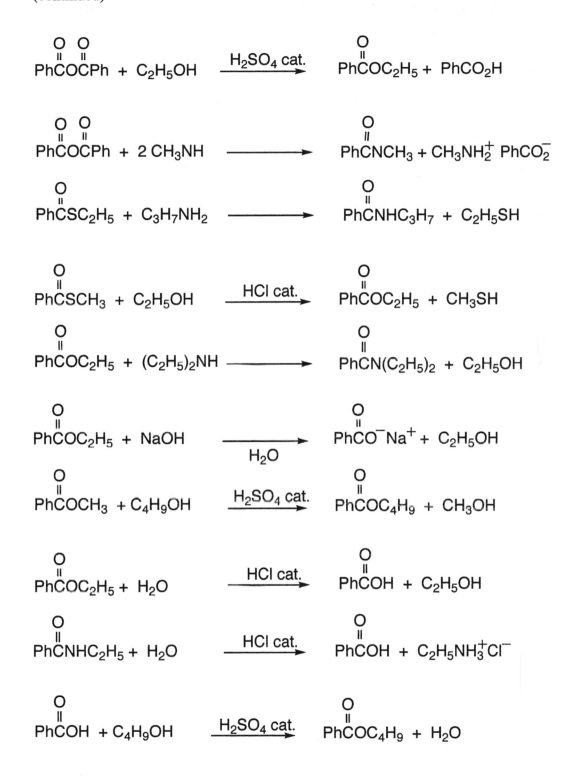

O   O
‖   ‖
PhCOCPh + C₂H₅OH $\xrightarrow{\text{H}_2\text{SO}_4 \text{ cat.}}$ PhCOC₂H₅ + PhCO₂H

O   O
‖   ‖
PhCOCPh + 2 CH₃NH $\longrightarrow$ PhCNCH₃ + CH₃NH₂⁺ PhCO₂⁻

O
‖
PhCSC₂H₅ + C₃H₇NH₂ $\longrightarrow$ PhCNHC₃H₇ + C₂H₅SH

O
‖
PhCSCH₃ + C₂H₅OH $\xrightarrow{\text{HCl cat.}}$ PhCOC₂H₅ + CH₃SH

O
‖
PhCOC₂H₅ + (C₂H₅)₂NH $\longrightarrow$ PhCN(C₂H₅)₂ + C₂H₅OH

O
‖
PhCOC₂H₅ + NaOH $\xrightarrow[\text{H}_2\text{O}]{}$ PhCO⁻Na⁺ + C₂H₅OH

O
‖
PhCOCH₃ + C₄H₉OH $\xrightarrow{\text{H}_2\text{SO}_4 \text{ cat.}}$ PhCOC₄H₉ + CH₃OH

O
‖
PhCOC₂H₅ + H₂O $\xrightarrow{\text{HCl cat.}}$ PhCOH + C₂H₅OH

O
‖
PhCNHC₂H₅ + H₂O $\xrightarrow{\text{HCl cat.}}$ PhCOH + C₂H₅NH₃⁺Cl⁻

O
‖
PhCOH + C₄H₉OH $\xrightarrow{\text{H}_2\text{SO}_4 \text{ cat.}}$ PhCOC₄H₉ + H₂O

2.   The general classes of acyl compounds in Problem 1 are ordered according to their relative reactivity with water. The IR carbonyl absorptions in the gas phase are as follows:

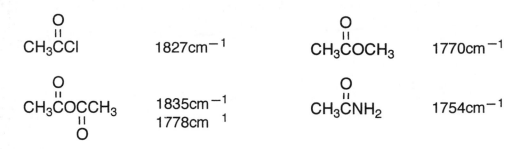

Analyze the structures of the different classes of carboxylic acid derivatives, and propose reasons for the observed relative reactivities and carbonyl stretching frequencies.

3.

   a.   Predict which of the following two competing reactions will be faster, and explain your choice:

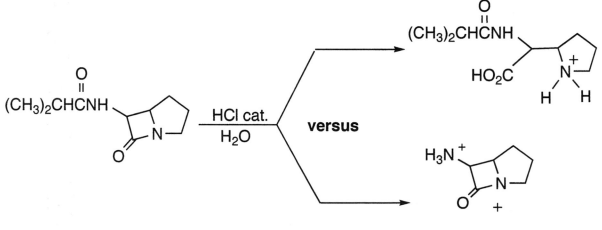

3.   (continued)

b.   The ester shown is labeled with oxygen-18 as indicated ($*O = {}^{18}O$). Give two possible mechanisms for the hydrolysis, and explain how one of the mechanisms is rigorously excluded by the labeling experiment.

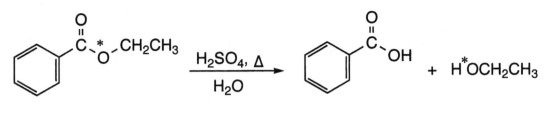

c.   Nitriles are at the same oxidation level as carboxylic acids and share the characteristic of a strongly polarized multiple bond. Propose a mechanism for the hydrolysis of a nitrile in aqueous acid to give the corresponding carboxylic acid.

$$RCN \xrightarrow[H_2O]{H_3O^+} RCO_2H$$

4.   Propose pathways to accomplish the transformations that follow. You may use other organic reactants as required.

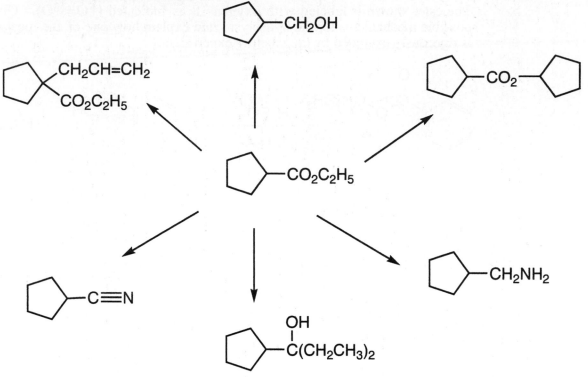

5. Derivatives of carboxylic acids are important in many technologies. Consider the following applications:

a. Benzyloxychloroformate (BO$_2$C–Cl), PhCH$_2$O$\overset{\overset{\text{O}}{\|}}{\text{C}}$Cl, is an important reagent used in the synthesis of peptides and proteins as a protective group for the amino groups of amino acids.

    i.   Show how BO$_2$C-Cl could be synthesized from phosgene.

    ii   Give a structure for the product of the reaction of BO$_2$C-Cl with the amino acid glycine, H$_2$NCH$_2$CO$_2$H.

b. Dacron® is a polyester formed by the transesterification of dimethyl terephthalate with ethylene glycol to give a long-chain polymer molecule. Provide a partial structure of Dacron that shows two of the repeating units of the polymer.

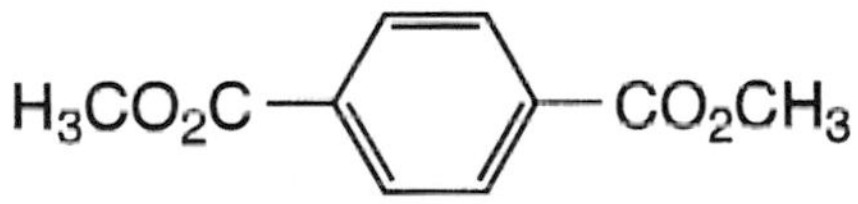

H$_3$CO$_2$C—⟨benzene⟩—CO$_2$CH$_3$ HO⁀OH

dimethyl terephthalate 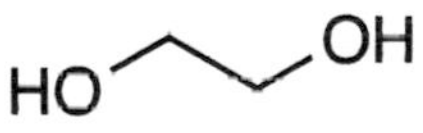   ethylene glycol

5.   (continued)

c.   Propose ways to synthesize the following compounds from the indicated starting
material:

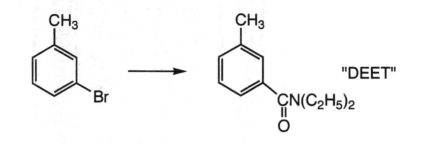

"DEET"

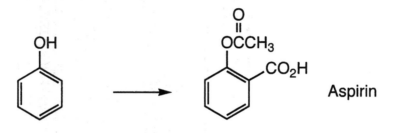

Aspirin

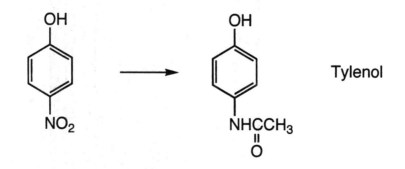

Tylenol

Reflection: Write a general mechanism that integrates the chemistry of nucleophilic addition and substitution reactions of aldehydes, ketones, carboxylic acids, and carboxylic acid derivatives. Construct a "reaction map" that connects the various structures to the mechanism by identifying the structural feature that determines the chemistry.

WORKSHOP 27

Lipids

Purpose: In this Workshop, you will explore the connections among the physical, chemical, and biological properties of lipids. You should be able to apply fundamental principles of structure and reactivity to develop a molecular-level understanding of the properties and function of this class of biomolecules.

Expectations: You should review, and be prepared for an initial discussion about, the meaning of the following key words and concepts: intermolecular interactions (see Workshop 3), Claisen condensation, *cis-trans* double-bond isomerization, and carbocation chemistry.

1. Geranyl pyrophosphate is converted to $\propto$-terpineol and/or limonene, depending on the organism. Pyrophosphate is a good leaving group. Produced biochemically, these terpene products are optically active. For example, R-(+)-limonene is found in the cones of the southern cypress tree, S-(-)-limonene is found in the needles of the Douglas fir tree, and both S-(-)-limonene and -$\propto$-terpineol occur in the essential oil obtained from the heartwood of the Douglas fir.

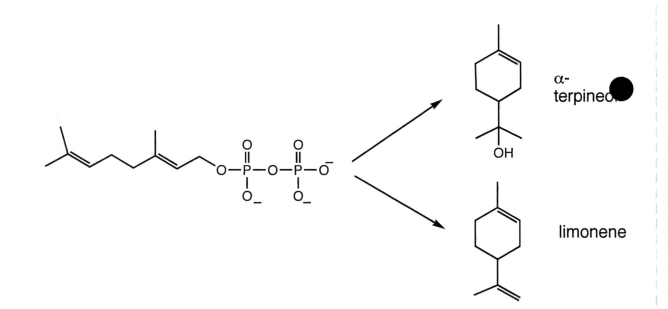

1. (continued)

       a.   Propose a simple reaction mechanism that can readily account for the formation of these terpene natural products.

       b.   Refer to your mechanism, and indicate the point at which the chirality is generated. Explain how a biological system can control the stereochemistry of this step.

2. The biochemical steps in the synthesis of long-chain fatty acids from acetate units are shown here for the formation of butanoate. The cycle can repeat, introducing two carbons per cycle, forming hexanoate, octanoate, etc. ("ACP-SH" is shorthand for "acetyl carrier protein.")

$$H_3O^+ + CH_3\overset{O}{\overset{\|}{C}}\text{-SACP} + {}^-O\text{-}\overset{O}{\overset{\|}{C}}\text{-CH}_2\overset{O}{\overset{\|}{C}}\text{-SACP} \rightleftharpoons CH_3\overset{O}{\overset{\|}{C}}CH_2\overset{O}{\overset{\|}{C}}\text{-SACP}$$

$$+ CO_2 + ACPSH + H_2O$$

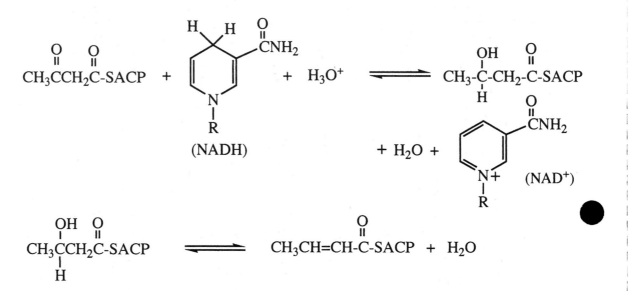

$$CH_3\overset{O}{\overset{\|}{C}}CH_2\overset{O}{\overset{\|}{C}}\text{-SACP} \; + \; \text{(NADH)} \; + \; H_3O^+ \rightleftharpoons CH_3\text{-}\underset{H}{\overset{OH}{\overset{|}{C}}}\text{-CH}_2\text{-}\overset{O}{\overset{\|}{C}}\text{-SACP}$$

$$+ \; H_2O \; + \; \text{(NAD}^+\text{)}$$

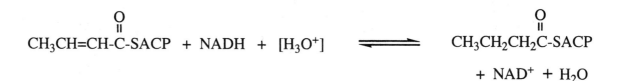

$$CH_3\underset{H}{\overset{OH}{\overset{|}{\underset{|}{C}}}}CH_2\overset{O}{\overset{\|}{C}}\text{-SACP} \rightleftharpoons CH_3CH=CH\text{-}\overset{O}{\overset{\|}{C}}\text{-SACP} \; + \; H_2O$$

$$CH_3CH=CH\text{-}\overset{O}{\overset{\|}{C}}\text{-SACP} \; + \; NADH \; + \; [H_3O^+] \rightleftharpoons CH_3CH_2CH_2\overset{O}{\overset{\|}{C}}\text{-SACP}$$

$$+ \; NAD^+ \; + \; H_2O$$

2.   (continued)

a.   Propose a reasonable general mechanism for each of the preceding reactions, and identify reactions you have previously studied that are analogous to these biochemical reactions. Use the curved-arrow formalism to show bond-making and bond-breaking processes.

b.   Show how the product of the final step can be used in the next stage of chain growth to make a C_6 product.

3. Consider the compartmentalized biological cellular system. The environment inside and outside the cell is aqueous. Relatively small molecules of low polarity, such as prescription drugs, pass through the cell membrane. In fact, it is important that an equilibrium concentration of a drug inside and outside the cell be established for the drug to be effective. The major components of the cell membrane are glycerolphospholipids, such as the phosphatidylcholine molecule shown next. The long-chain fatty acid esters may be saturated or may contain *cis*-double bonds. Review how the salts of long-chain fatty acids aggregate in water (Workshop 3, Problem 5b). The phosphatidylcholine molecules are similar in that they are amphipathic; that is, they have nonpolar (hydrophobic) tails and polar (hydrophilic) head groups. However, being different molecules, they aggregate somewhat differently to form the cell membrane.

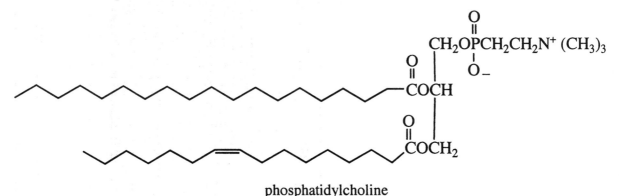

phosphatidylcholine

a. Experiment with different aggregations of phosphatidylcholine molecules to find an arrangement that allows the membrane to interact with the aqueous environments inside and outside the cell and, at the same time, keep these two environments separated.

3.   (continued)

b.   For a prescription drug to be effective, it must possess at least limited water solubility and be able to pass by diffusion through the cell wall made up of glycerolphospholipids. A simple test model to see if a drug meets these requirements involves measuring the distribution coefficient between 1-octanol and aqueous solution buffered at a pH of 7.4. If the distribution coefficient octanol/water = 1.5–3, the drug molecule is judged to have the appropriate solubility properties. Explain why a distribution coefficient of 1.5–3 is a good predictor for a drug molecule to be capable of being transported to the cell and, through its membrane, into the cell. Also, explain why distribution coefficients outside this range are undesirable.

4. *trans*-Fats are implicated in increases in low-density lipid (LDL) levels in the blood. As a result, the Food and Drug Administration is requiring manufacturers to disclose the *trans*-fat content of all food products by 2006.

Crisco and margarines are made from oils obtained from plants; soybean oil is the most common starting material. The naturally occurring oils are mixtures of triglycerides, triesters of fatty acids, and glycerol. The majority of the acid components of the tryiglycerides are unsaturated fatty acids containing *cis*-double bonds. Because of the *cis* geometry in the side chains, the molecules do not pack together well; the result is a low-melting compound, an oil at room temperature. Solid products such as Crisco and margarines are made by catalytic hydrogenation of the *cis*-double bonds. The resulting alkyl chains pack together better; the stronger inter- and intramolecular forces give a product with a higher melting point. Unfortunately, the catalytic hydrogenation process isomerizes some of the *cis*-double bonds to *trans*-double bonds. This is the source of the *trans*-fats.

One possibility is that the isomerization is directly related to the mechanism of the hydrogenation. Consider two possible mechanisms for catalytic hydrogenation. Both mechanisms start with the reaction of hydrogen with the catalytic metal:

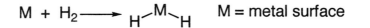

Mechanism A: concerted

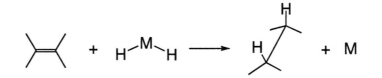

Mechanism B: stepwise

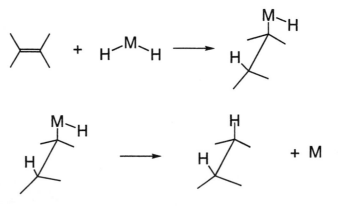

4.   (continued)

a.   Give representations of the structures of a triglyceride of three saturated fatty acids and a triglyceride containing two saturated fatty acids esterified at C_1 and C_3 and one *cis-* monounsaturated fatty acid esterified at C-2 of glycerol. Explain the differences in inter- and intramolecular forces that lead to different melting points.

b.   Explain how one or the other of the proposed mechanisms could accommodate the observed *cis-/trans-* isomerization that occurs during catalytic hydrogenation.

A different possibility is that the isomerization is the result of a reaction of the catalytic metal surface with the *cis*-alkene.

c.   Propose an experimental test to distinguish this hypothesis from the hydrogenation hypothesis.

Reflection: List the key reactions or ideas from (nonbiological) organic chemistry that helped you understand the (biological) reactions and properties of lipids.

WORKSHOP 28

Amines

Purpose: Amines are organic derivatives of ammonia. They are important items of commerce and essential components of living systems. You already know their most important characteristics: basicity and nucleophilicity. This Workshop provides a short refresher course and the opportunity to consider some new reactions to prepare or utilize amines.

Expectations: You should be prepared to discuss basicity; proton affinity; nucleophilicity; nucleophilic additions and substitutions; primary, secondary, tertiary, and quaternary nomenclature; and reductive amination. You should also be prepared to use your book to look up reactions that puzzle you.

1.   a.   The compounds that follow are stored in two unlabeled bottles. Both compounds are insoluble in water. Using a simple solubility test, how would you determine which bottle contains which compound? Give the chemical reaction for the positive test. Which spectroscopic method could be used to distinguish the two compounds? Explain exactly what you would do and exactly what you would expect to observe for the two different compounds in each test.

    b.   Draw the structure of Novocain™, is which is the salt formed from procaine (structure shown) and one equivalent of hydrogen chloride. Explain the basis of your choice.

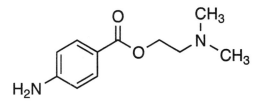

227

2.   a.   Explain, verbally and with structures, the conceptual basis of the separation of a racemic mixture into its enantiomeric components.

b.   Draw flowcharts to show two different ways that you could use an optically active amine, $(+)-RNH_2$, to resolve a racemic mixture of an acid, $(\pm)-RCOOH$. One of your methods should involve different rates of reaction. Explain your choice of reagents and reaction conditions at each step.

$$(+) - RNH_2 = (S) - C_6H_5CH_2\underset{\underset{\displaystyle CH_3}{|}}{C}HNH_2$$

3. Give structures for compounds **S** through **W**, and specify missing reagents and conditions in the following scheme:

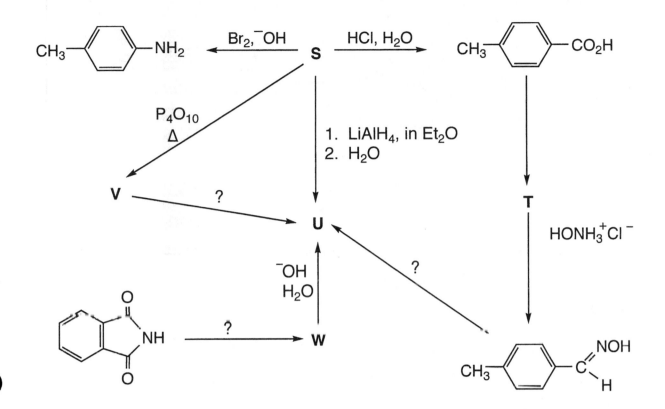

4.   Give the products of the reactions that follow. Use the curved-arrow formalism to show
      the bond-making and bond-breaking processes. Show stereochemistry as appropriate.

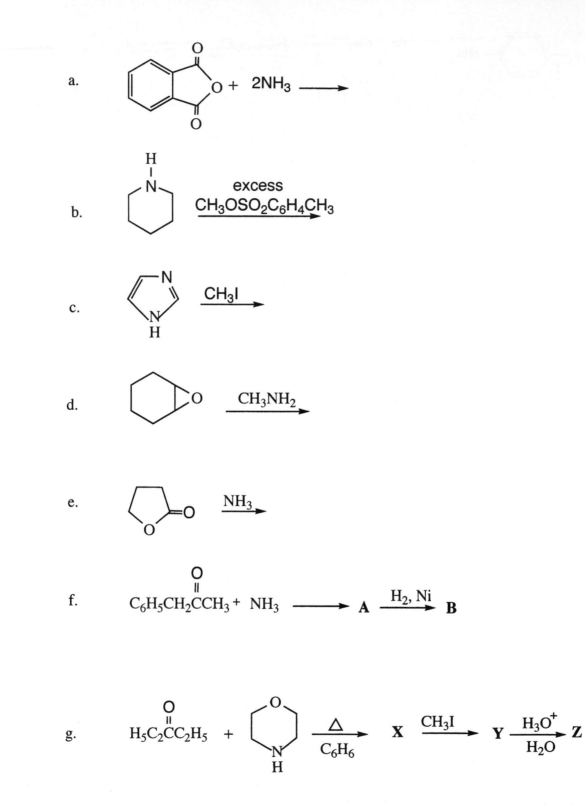

a.

b.

c.

d.

e.

f.

g.

5.   Consider the following observations, and propose a mechanism for the formation of the β-aminoketone:

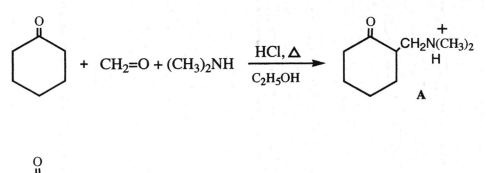

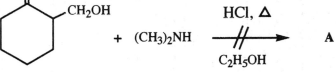

Reflection: Work with your colleagues to prepare a reaction map that shows ways to prepare amines and characteristic reactions of amines. Try to arrange your map so that it groups together reactions which have a common mechanistic theme.

WORKSHOP 29

Amino Acids and Peptides

Purpose: In Workshop 16, we considered the special properties of conjugated systems—two or more functional groups with interacting p-orbitals. Amino acids are difunctional molecules (as the name announces) in which the functional groups are *not* conjugated. Nevertheless, the amino acids present some unique properties that are more than just the sum of the properties of the carboxylic acid and amino functional groups. Among these unique properties are the amino acid's amphoteric acid–base properties and intermolecular bond formation to form polymers (peptides). Since amino acids and peptides are essential components of living systems, this Workshop focuses on making the transition from organic to bioorganic chemistry.

Expectations: You should review the pK_a of carboxylic acids and ammonium ions, preparations and reactions of amides, protecting and activating groups, and the specificity of enzymatic hydrolysis.

1.   Identify the expected state of ionization and the net charge on the following amino acids and peptides at physiological pH (7.3). This is a good problem to do in Round Robin format; explain your choices to your colleagues.

     a.   Ala; His; Tyr; Lys; Glu; Cys

1.   (continued)

     b.   isoleucylarginine; polyglycine; AlaTrpGluLeu

2.   Sketch out a titration curve (pH vs. equivalents of added ⁻OH) for the titration of glycine hydrochloride. Specify the major species in solution at pH 1, 7, and 12.

3. Simple peptides have a regular structure in which the four-atom, three-bond unit is planar:

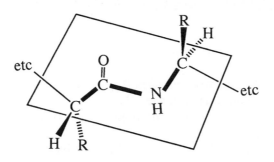

Explain the origins of the planar, *trans* geometry for the peptide bond. (Hint: Remember that the amide carbonyl absorbs in the IR at 1680 cm^{-1}.) Also, explain the regular alternation of the R groups, behind the plane, in front of the plane, etc.

4.   Consider an unknown heptapeptide consisting of the following amino acids:

Arg, Asp, His, Ile, Phe, Tyr, and Val

Deduce the primary structure of the peptide from the following data:

- Edman degradation releases Asp.
- Carboxylpeptidase releases Phe.
- Chymotrypsin cleaves the peptide to a tripeptide and a tetrapeptide, **A** and **B**, respectively.
- **A** has Ile at its N-terminus and Phe at its C-terminus.
- **B** can be partially hydrolyzed to give dipeptides ArgVal, AspArg and ValTyr.

(Do not be afraid to use your book if you cannot remember specific cleavages.)

OBSERVATION   DEDUCTION

5. Dicyclohexylcarbodimide (DCC) is used to facilitate the formation of a peptide bond:

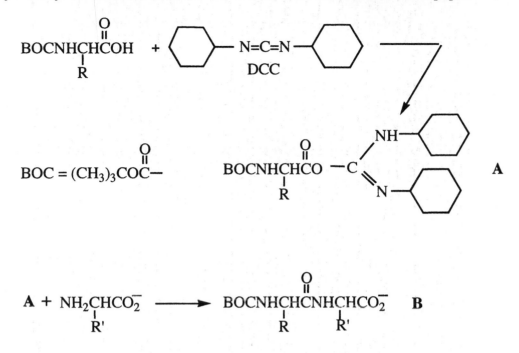

a. Give an arrow-pushing mechanism for the formation of **A** and its subsequent reaction to give **B**. Explain how DCC activates one amino acid for reaction with the other amino acid to make the peptide bond in **B**.

5.   (continued)

b.   Explain why the starting amino acid must be protected at the amino group.

c.   Explain why the DCC procedure is especially good for the repeated sequential formation of polypeptides.

6.   Chymotrypsin catalyzes the hydrolysis of the peptide bonds of the aromatic amino acids Phe, Tyr, and Trp.

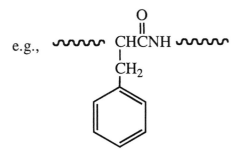

Re-create the chymotrypsin template on the board, and work with your teammates to follow the chymotrypsin-catalyzed hydrolysis of the peptide shown in the preceding diagram.

6.   (continued)

The Chymotrypsin Active Site

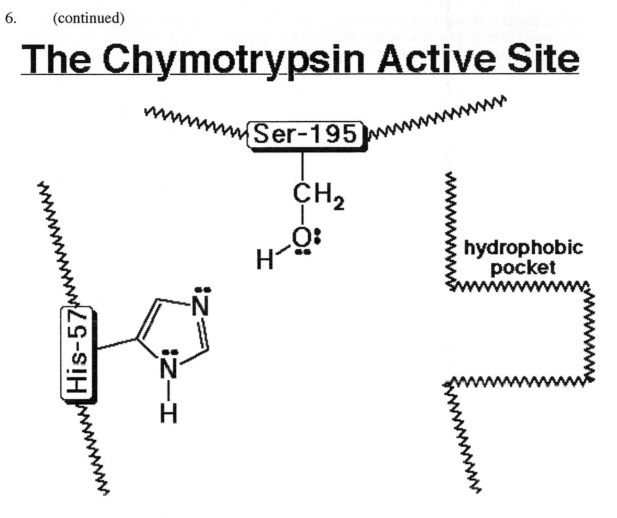

(1)   A polypeptide enters the active site and the aromatic amino acid side chain and is recognized by the hydrophobic pocket.

(2)   Ser-195 acts as a nucleophile to add to the carbonyl group of the peptide, while His-57 removes a proton from Ser-195 to make it a better nucleophile.

(3)   The tetrahedral intermediate eliminates an amino group, while His-57 donates a proton to the amino group to make it a better leaving group.

(4)   The broken peptide fragment (the N-part) leaves the active site and is replaced by water.

(5)   Water acts as a nucleophile to add to the carbonyl group of the acyl enzyme intermediate, while His-57 removes a proton from water to make it a better nucleophile.

(6)   The tetrahedral intermediate eliminates the Ser-195 group, while His-57 donates a proton to Ser-195 to make it a better leaving group.

(7)   The other part of the broken peptide (the C-part) leaves the active site and is replaced by another polypeptide (as in step 1) or by a polypeptide fragment of the original.

Reflection: Work with your colleagues to find how many different peptides can be constructed from three amino acids. Write a short paragraph explaining why the multiplicity of peptide structures is important for living systems.

WORKSHOP 30

Metabolism

Purpose: It is wonderfully satisfying to find that biology runs on organic reactions. In this Workshop, you will apply your understanding of organic reactions to biological examples. The glycolysis pathway oxidizes glucose to carbon dioxide. The last 10 reactions in the glycolysis pathway are known as the Krebs cycle. You will explore the reactions of the Krebs cycle, with emphasis on the relationships between the biological reactions and their organic counterparts Generally, the reactions are familiar organic reactions, but with some new approaches required in order to run under biological conditions.

Expectations: You should review and be prepared for an initial discussion about the meaning of the following key words and concepts: (a) cofactors $NAD^+/NADH$ and $FAD/FADH_2$, (b) coenzyme A, and (c) ATP/ADP.

The first part of the glycolysis pathway converts glucose to pyruvate:

$$C_6H_{12}O_6 + 2NAD^+ + 4B: \longrightarrow 2\ CH_3COCO_2^- + 4BH^+ + 2NADH$$

Pyruvate is subsequently converted to acetyl CoA, the starting material for the Krebs cycle (a.k.a. the citric acid cycle or tricarboxylic acid cycle):

$$2\ CH_3COCO_2^- + 2NAD^+ + 2CoASH \longrightarrow 2CO_2 + 2NADH + 2CH_3COSCoA$$

1. Distribute the 10 steps in the Krebs cycle among the members of the group, with each member taking responsibility for 1 or 2 steps. Make notes on the scheme to characterize each of the 10 *transformations* according to the following instructions:

   a. Write a balanced equation and describe the transformation with as much specificity as you can (e.g., oxidation of a secondary alcohol to a ketone).

   b. Indicate the steps that require a cofactor to accomplish an oxidation. Use $NAD^+/NADH$ to interconvert alcohols and carbonyl groups, and use $FAD/FADH_2$ to interconvert hydrocarbons and alkenes.

   c. Indicate the steps in which stereocenters are formed.

   d. Explain why each of these stereocenters is formed with absolute enantiomeric specificity.

   e. Explain what is accomplished by steps 2 and 3 and why this change is essential for the overall scheme.

   f. Specify reagents and conditions that could be used to accomplish the same conversion in the organic chemistry laboratory.

The Krebs Cycle

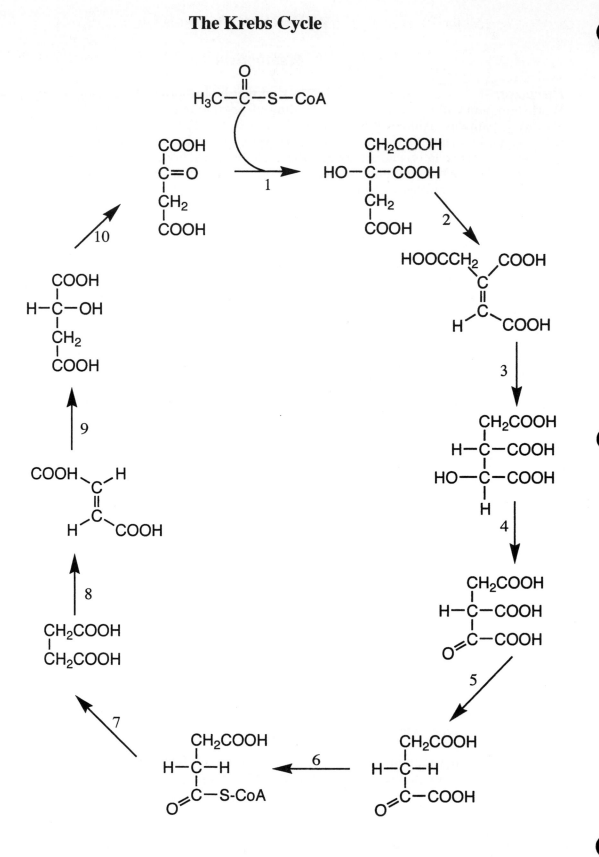

2.   Consider the 10 reactions of the Krebs cycle, *all together*.

    a.   Write a balanced reaction for the entire cycle, taking particular care to add up the cofactors that are used and generated.

    b.   In other oxidative phosphorylation reaction sequences, each equivalent of NADH ultimately creates 3 equivalents of ATP and each equivalent of $FADH_2$ creates 2 equivalents of ATP. In addition, the hydrolysis of the thioester bond in step 7 is coupled to the formation of 1 equivalent of ATP.

        i.   How many ATP's can be created from an acetyl CoA?
        ii   How many from one glucose?

3.   The metabolism of fats and proteins also feeds acetyl CoA into the Krebs cycle. In the first step of the catabolism of fats, the esters are hydrolyzed to the component fatty acids and glycerol.

Consider the structural relationship of succinic acid (in the Krebs cycle) and a fatty acid such as palmitic acid:

$$CH_3(CH_2)_{12}CH_2CH_2CO_2H \qquad \text{Palmitic Acid}$$

a.   Propose a series of intermediates for the metabolism of palmitic acid to acetyl CoA and myristic acid, $CH_3(CH_2)_{12}CO_2H$.

b.   Analyze the relationship of your proposal to the mechanism describing the biosynthesis of fatty acids.

4. The cofactor nicotinamide adenine dinucleotide (NAD⁺) and its redox partner NADH have complicated structures. However, the redox part of the molecules is the pyridinium/dihydropyridine half-cell.

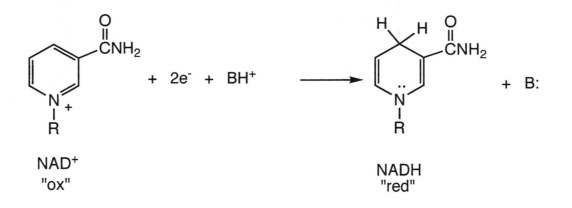

NAD⁺
"ox"

NADH
"red"

Consider the reverse reaction in order to understand the mechanism of reaction of this important oxidizing agent in the Krebs cycle:

$$R_2C = O + NADH + BH^+ \longrightarrow R_2CHOH + NAD^+ + B:$$

a. Identify two nonbiological analogs of NADH. Use curved arrows to show how the nonbiological and biological reagents react by comparable mechanisms.

b. Using the principle of microscopic reversibility, write a mechanism for the oxidation of (S)-malate to oxaloacetate by NAD⁺ (step 10 in the Krebs cycle). Be sure to use the curved-arrow formation to show the bond-making and bond-breaking processes.

Reflection: Discuss the ways in which biological reactions differ from similar reactions that could be carried out in the organic chemistry laboratory (e.g., oxidation of a secondary alcohol to a ketone). Use specific examples from the Krebs cycle. Explain the constraints inherent in a biological reaction and the constraints inherent in an organic chemistry laboratory reaction. Discuss some of the tactics used in the Krebs cycle to get around the constraints of biological systems.

REVIEW

1. Make lists showing as many examples as you know of reactions that involve each of the concepts that follow. It may be fun to do this problem in a round-robin format.

   a. Formation of new C–C bonds

   b. Carbanion intermediates

   c. Carbocation intermediates

   d. Reduction of organic molecules

   e. Oxidation of organic molecules

   f. Electrophilic addition to π bonds

   g. Nucleophilic addition to π bonds

   h. Concerted mechanisms

2. Consider the following problems about equilibria and rates:

   a. Rationalize the observed equilibria:

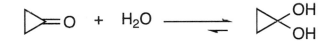

$$\triangleright\!\!=\!\!O \ + \ H_2O \ \rightleftharpoons \ \triangleright\!\!<\begin{smallmatrix}OH\\OH\end{smallmatrix}$$

versus

$$\triangleright\!\!=\!\!O \ + \ H_2O \ \rightleftharpoons \ \triangleright\!\!<\begin{smallmatrix}OH\\OH\end{smallmatrix}$$

$$\underset{\substack{\parallel\\O}}{CH_3CH} \ + \ H_2O \ \rightleftharpoons \ CH_3CH\!\!<\begin{smallmatrix}OH\\OH\end{smallmatrix}$$

versus

$$\underset{\substack{\parallel\\O}}{Cl_3CCH} \ + \ H_2O \ \rightleftharpoons \ Cl_3CCH\!\!<\begin{smallmatrix}OH\\OH\end{smallmatrix}$$

2.   (continued)

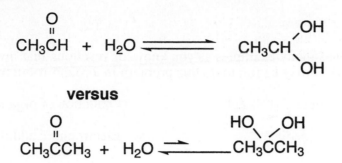

versus

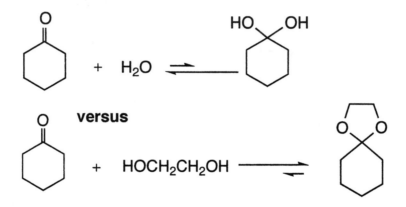

versus

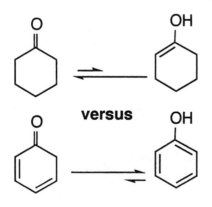

versus

248

2.   (continued)

b.   For each of the pairs that follow, circle the compound that corresponds to the faster rate of exchange with $H_2^{18}O$ in the presence of 0.01M $H_3^{18}O^+$. Explain your choices.

$$\underset{\displaystyle CH_3CCH_3}{\overset{\displaystyle \overset{O}{\|}}{}}$$ **versus**

$$\underset{\displaystyle CH_3CH_2CH}{\overset{\displaystyle \overset{O}{\|}}{}}$$ **versus** $H_2C = O$

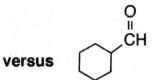

versus

2.   (continued)

  c.   Arrange the following molecules in order of decreasing rate of reactivity with
          0.1M NaOH solution. Explain your choices.

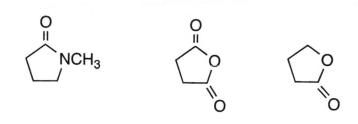

  d.   Reaction of the compound shown next with iodomethane led to the formation of
          two isomeric products, **A** and **B**. When each of these was first treated with
          Ag_2O/H_2O and then heated, a mixture of compounds **C** and **D** was formed.
          Compound **C** could be resolved into enantiomers, but **D** could not.

Ph

N

(Note: $2\ R_4N^+\ I^- + Ag_2O \xrightarrow[H_2O]{} 2AgI + 2R_4N^+HO^-$)

      i.   Propose structures for **A** and **B**, and name the structural relationship
              between them.
      ii.   Propose structures for **C** and **D**. Predict the approximate product
              composition: **C/D** $\gg$ 1; **C/D** $\ll$ 1; or **C/D** ~1.

3.   Consider each of the reactions that follow. First, decide whether the reaction is net oxidation, net reduction, or neither. Second, specify an appropriate reagent to bring about the transformation. Specify the half-cell for the redox reagents.

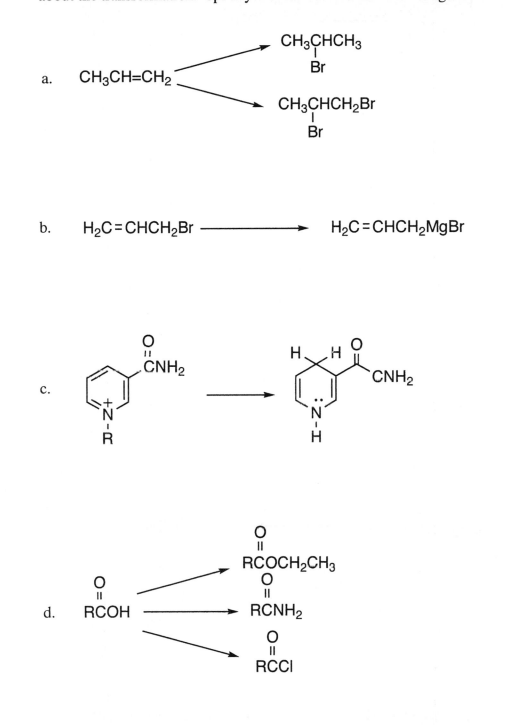

3   (continued)

e.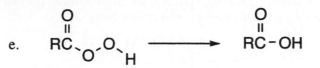

f.

g.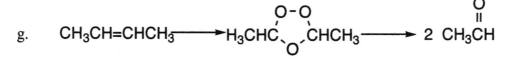

4.   The reactions that follow do not occur, or occur only slowly, under ordinary conditions in the absence of a catalyst. Suggest the necessary catalyst that makes these reactions useful. For each reaction, indicate the new mechanistic pathway (with lower activation energy) that is opened up by the catalyst.

a. $CH_2=CH_2 + H_2 \longrightarrow H_3C-CH_3$

b. $CH_2=CH_2 + H_2O \longrightarrow CH_3CH_2OH$

c. $CH_3C\equiv C-H + H_2O \longrightarrow H_3CCCH_3$ (with C=O)

d. $(CH_3)_3CH + Br_2 \longrightarrow (CH_3)_3CBr + HBr$

4. (continued)

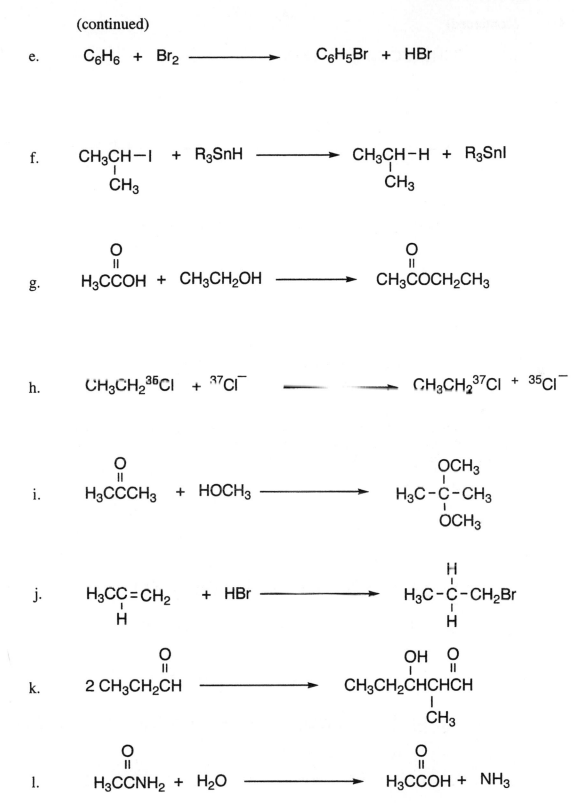

e. C_6H_6 + Br_2 $\longrightarrow$ C_6H_5Br + HBr

f. CH_3CH-I + R_3SnH $\longrightarrow$ CH_3CH-H + R_3SnI
      | |
      CH_3 CH_3

g. H_3CCOH + CH_3CH_2OH $\longrightarrow$ $CH_3COCH_2CH_3$ (with C=O)

h. $CH_3CH_2{}^{35}Cl$ + ${}^{37}Cl^-$ $\longrightarrow$ $CH_3CH_2{}^{37}Cl$ + ${}^{35}Cl^-$

i. H_3CCCH_3 (C=O) + $HOCH_3$ $\longrightarrow$ $H_3C-C-CH_3$ with OCH_3 above and OCH_3 below

j. $H_3CC=CH_2$ + HBr $\longrightarrow$ $H_3C-C-CH_2Br$ with H above and H below
      |
      H

k. $2\ CH_3CH_2CH$ (C=O) $\longrightarrow$ $CH_3CH_2CHCHCH$ with OH and C=O, and CH_3 below

l. H_3CCNH_2 (C=O) + H_2O $\longrightarrow$ H_3CCOH (C=O) + NH_3

253

4.   (continued)

m. $(CH_3)_3COC_6H_5$ + H_2O ⟶ $(CH_3)_2C=CH_2$ + HOC_6H_5

n. C_6H_6 + C_6H_5COCl ⟶ $C_6H_5COC_6H_5$ + HCl

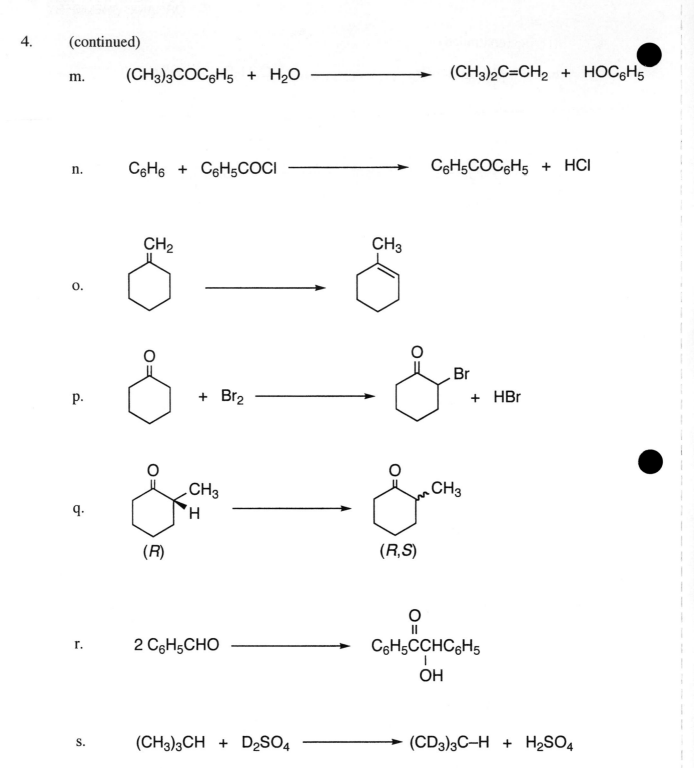

o.

p. + Br_2 ⟶ + HBr

q.   (R)   (R,S)

r.   2 C_6H_5CHO ⟶ $C_6H_5CCHC_6H_5$ (with =O and OH)

s. $(CH_3)_3CH$ + D_2SO_4 ⟶ $(CD_3)_3C–H$ + H_2SO_4

4.   (continued)

t. $C_6H_5C\equiv CC_6H_5$ + H_2 ⟶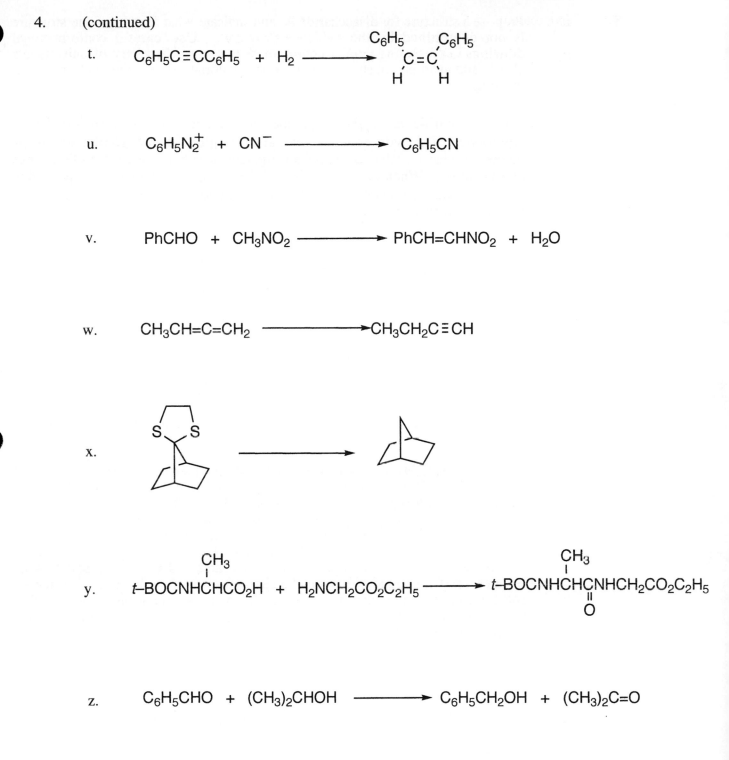

u. $C_6H_5N_2^+$ + CN^- ⟶ C_6H_5CN

v.   PhCHO + CH_3NO_2 ⟶ $PhCH=CHNO_2$ + H_2O

w. $CH_3CH=C=CH_2$ ⟶ $CH_3CH_2C\equiv CH$

x.

y. t–BOCNHCHCO$_2$H + $H_2NCH_2CO_2C_2H_5$ ⟶ t–BOCNHCHCNHCH$_2$CO$_2$C$_2$H$_5$

z. C_6H_5CHO + $(CH_3)_2CHOH$ ⟶ $C_6H_5CH_2OH$ + $(CH_3)_2C=O$

5.   a.   Propose a structure for disaccharide **A**, and indicate what feature of the structure is not determined by the evidence provided. Use careful conformational drawings in presenting your structure; mark the stereochemistry of substituents clearly and unambiguously as axial (ax) or equatorial (eq) on the conformational drawing.

An disaccharide **A**, $C_{11}H_{20}O_{10}$, does *not* react with $Ag(NH_3)_2^+$ or Cu^{2+}. Hydrolysis of **A** with aqueous acid gives D-glucose and D-ribose. A β-glucosidase has no effect on **A**, but an α-glucosidase hydrolyzes **A** to D-glucose and D-ribose. When **A** is methylated [$(CH_3O)_2SO_2$, NaOH], a heptamethyl ether **B** is obtained that can be hydrolyzed with aqueous acid to 2,3,4,6-tetra-O-methyl-D-glucose and 2,3,4-tri-O-methyl-D-ribose.

    b.   Provide structures for **A**, **B**, and **C** that are consistent with the following observations:

Compound **A**, $C_7H_{15}NO$, was neutral and exhibited strong absorption in the IR spectrum at 1650 cm^{-1}, but none in the 3200–3400-cm^{-1} region. Extended heating of **A** with H_3O^+/H_2O gave compound **B** and, after the addition of NaOH, a nitrogen-containing compound **C**. **B** exhibited strong absorption in the IR spectrum at 1712 cm^{-1} and a very broad absorption over the range 2500–3500 cm^{-1}. The ^1H NMR spectrum of **B** consisted of the following absorptions: δ 1.20 (d, rel. area 6, $J = 7.0$ Hz), 2.4 (septet, rel. area 1, $J = 7.0$ Hz), and 12.4 (s, rel. area 1). Compound **C** exhibited a sharp absorption at 3335 cm^{-1} in the IR spectrum and reacted with nitrous acid at 0 °C to give a yellow oil. Compound **C** was synthesized by treating ethyl amine with (1) formic acetic anhydride and (2) $LiAlH_4$ (H_3O^+ workup).

5.   (continued)

c.   Give structures for **B** through **E** that are consistent with the following information:

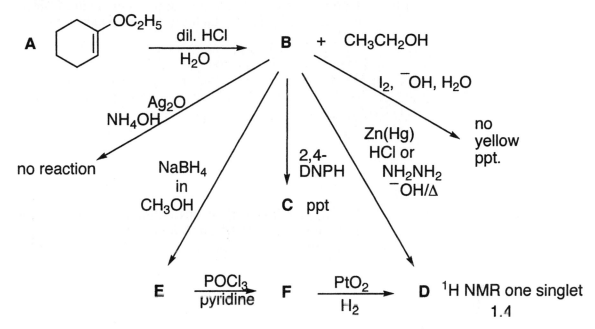

5.   (continued)

   d.   Provide structures that satisfy the following descriptions:

        i.   Upon vigorous oxidation with permanganate, C_8H_{10} is converted to a compound, $C_8H_6O_4$, that loses water upon heating.

        ii. $C_6H_{10}O_3$ exhibits strong absorption in the infrared spectrum at 1754 cm^{-1} and 1818 cm^{-1} and also exhibits the following H-NMR spectrum: 1.2 ppm (t, rel. area 3, $J = 7.0$ Hz) and 2.5 ppm (q, rel. area 2, $J = 7.0$ Hz).

        iii.   the stereoisomer of 3-hydroxycyclohexanecarboxylic acid that forms a lactone.

5.   (continued)

e.   Provide structures for the lettered compounds in the following reaction schemes:

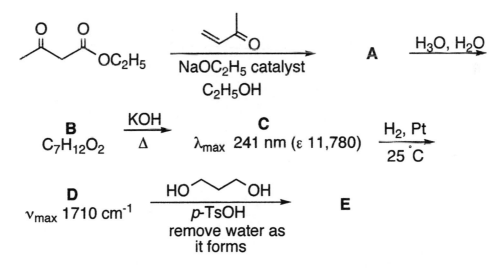

f.   Provide structures for the lettered compounds.

6. Provide structures for the major product(s) of the reactions shown next. Explain, verbally and with structures, your choice of the major product.

a. 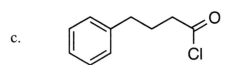 $\xrightarrow[\text{FeCl}_3]{\text{Cl}_2}$

b. $\xrightarrow[\substack{\text{AlCl}_3,\ \text{CuCl}\\ 2.\ \text{H}_2\text{O}}]{1.\ \text{CO, HCl}}$

c. 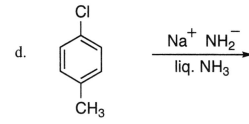 $\xrightarrow[2.\ \text{H}_2\text{O}]{1.\ \text{AlCl}_3\ (1.1\ \text{equiv})}$ $(\text{C}_{10}\text{H}_{10}\text{O})$

d. $\xrightarrow[\text{liq. NH}_3]{\text{Na}^+\ \text{NH}_2^-}$

6.   (continued)

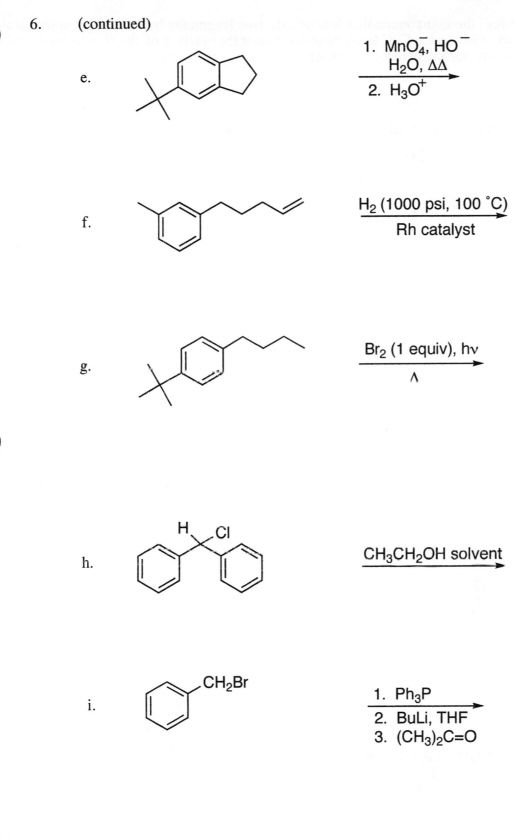

e. $\begin{array}{l} \text{1. MnO}_4^-, \text{HO}^- \\ \quad \text{H}_2\text{O, } \Delta\Delta \\ \hline \text{2. H}_3\text{O}^+ \end{array}$

f. $\dfrac{\text{H}_2 \text{ (1000 psi, 100 }^\circ\text{C)}}{\text{Rh catalyst}}$

g. $\dfrac{\text{Br}_2 \text{ (1 equiv), h}\nu}{\Lambda}$

h. $\text{CH}_3\text{CH}_2\text{OH solvent}$

i.   1. Ph_3P
       2. BuLi, THF
       3. $(\text{CH}_3)_2\text{C=O}$

7.   "Disconnect" the compounds that follow into two fragments by breaking appropriate C–C bonds. Choose the C–C bond carefully so that the reaction of the fragments would reassemble the starting compound. Specify the reactive fragments.

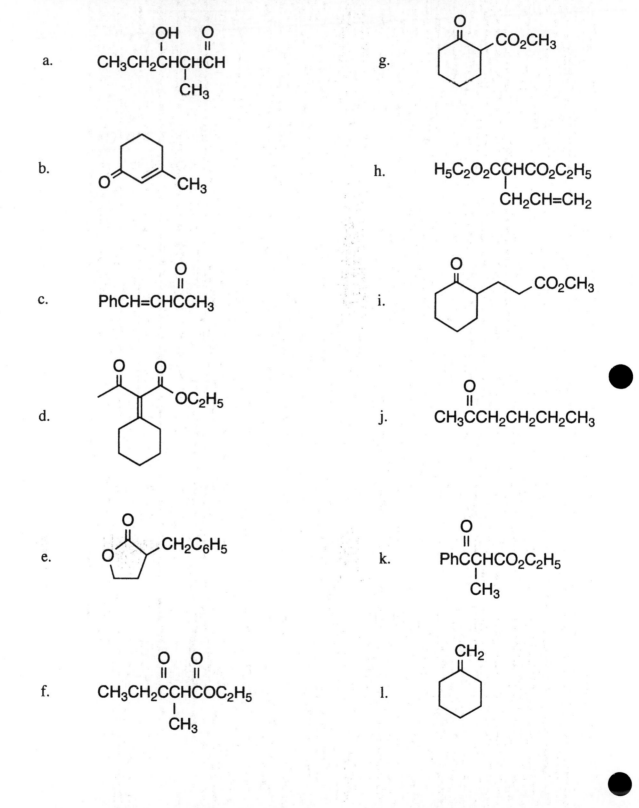

a.

$$\underset{\underset{CH_3}{|}}{CH_3CH_2CH{-}CHCH}$$

with OH on the CHCH₃ carbon and O (=O) on the terminal CH

g.

b.

c. $PhCH{=}CHCCH_3$ (with O double bond)

d.

e.

f. $\underset{\underset{CH_3}{|}}{CH_3CH_2CCHCOC_2H_5}$ (with two C=O)

h. $\underset{\underset{CH_2CH{=}CH_2}{|}}{H_5C_2O_2CCHCO_2C_2H_5}$

i.

j. $CH_3CCH_2CH_2CH_2CH_3$ (with C=O)

k. $\underset{\underset{CH_3}{|}}{PhCCHCO_2C_2H_5}$ (with C=O)

l.

262

8.   Show how to carry out the indicated synthetic conversions in an efficient manner. If more than one step is required, show the products for each step. Make as many synthetic connections as possible with compounds on the periphery of the "starburst."

a.

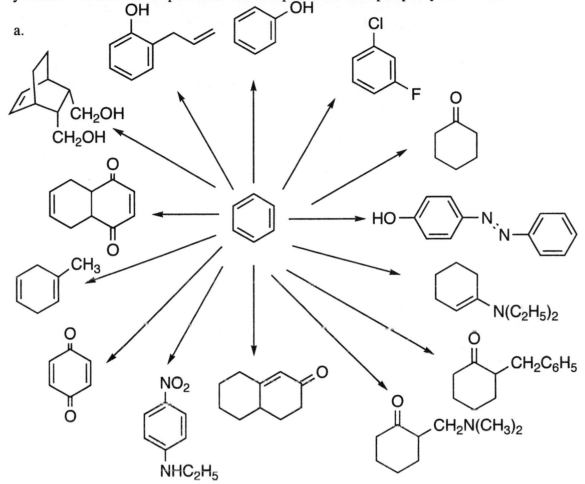

8.   (continued)

b.

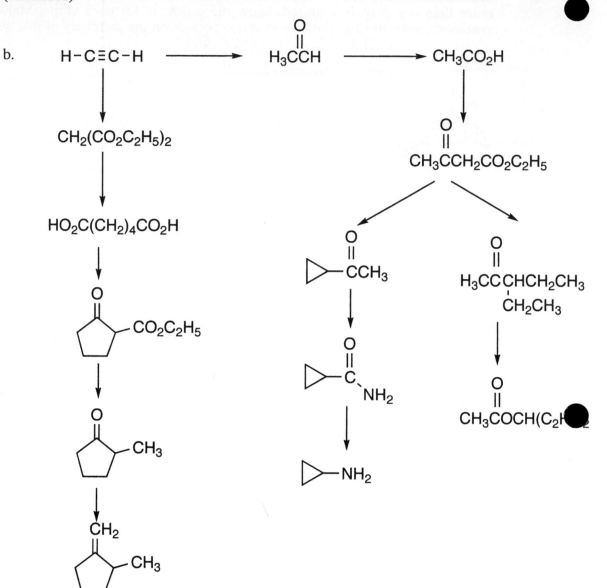

8   (continued)

c.

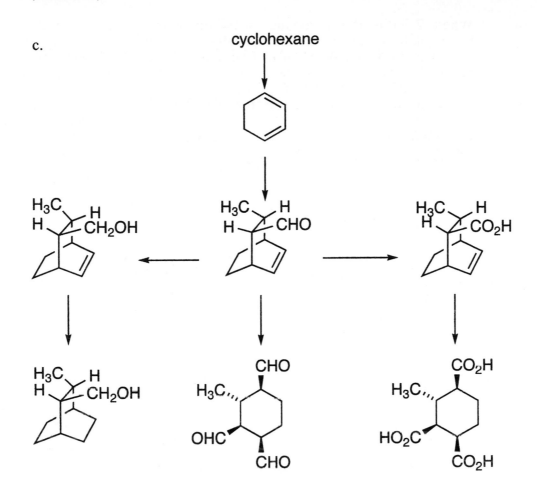

cyclohexane

9. Consider the following mechanistic puzzles:

a. When 2,2-dimethoxypropane is mixed with *n*-butanol and some *p*-toluenesulfonic acid in benzene, the equilibrium shown next is established. The lowest-boiling component of this mixture is methanol. Provide a reasonable mechanism for the reaction, and suggest an experimental technique that could be used to shift the equilibrium to the right. (*Note*: There is no water present; therefore, the reaction does **NOT** proceed by hydrolysis of the ketal to acetone.)

$$\underset{OCH_3}{\overset{OCH_3}{\diagdown\diagup}} + 2 \diagup\!\!\diagdown\!\!\diagup OH \xrightarrow[\text{benzene solvent}]{\textit{p}\text{-TsOH}} \diagup\!\!\diagdown\!\!\diagup + 2\ CH_3OH$$

b. Give reasonable mechanisms for each of the following reactions, clearly showing all important intermediates:

i. [structure: cyclopentanone N-phenyl imine] $+\ CH_3OH$ (excess) $\xrightarrow{\text{dry HCl}}$ [structure: 1,1-dimethoxycyclopentane] $+\ PhNH_3{}^+Cl^-$

ii. $BrCH_2CH_2CH_2CH_2$ [structure: cyclopentanone with side chain] $\xrightarrow[\text{2. Me}_3\text{COK}]{\text{1. Ph}_3\text{P}}$ [structure: bicyclic alkene]

9.   (continued)

c.   Pinacol rearrangement is a reaction that produces carbonyl compounds. Give a mechanism for the following reaction:

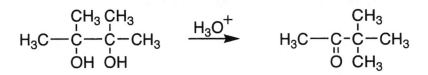

The following reaction gives only one product as shown.

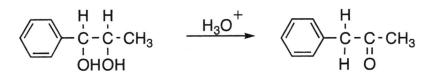

What are the other three possible products? Explain the observed selectivity.

9.   (continued)

d.   Give reasonable mechanistic pathways that account for the formation of two products under different reaction conditions. Explain why each is the preferred product under the reaction conditions specified.

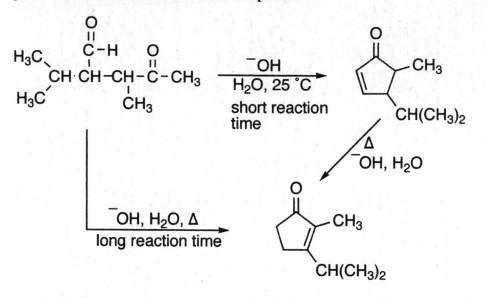

9.   (continued)

e.   Propose plausible mechanisms for each of the reactions that follow. Provide all
      important Lewis structures that contribute to the resonance hybrid for delocalized
      intermediates, and point out any important driving forces.

i.

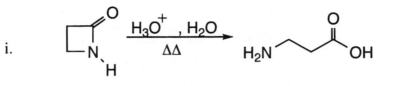

ii.

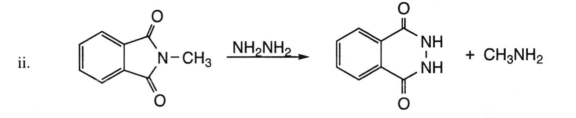

iii.

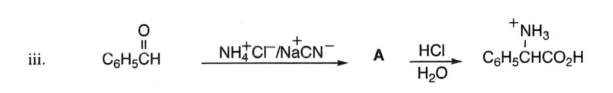

iv.

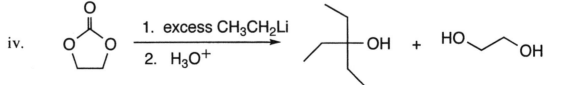

9.   (continued)

f.   The hydrolysis of esters is catalyzed by acid or base. The reaction can be carried out in isotopically labeled water, and the products can be analyzed for the presence of the label.

$$CH_3COR \xrightarrow[H_2O^{18}]{H_3O^+ \text{ or } HO^-}$$

(with the structure showing $CH_3\overset{\displaystyle O}{\overset{\|}{C}}OR$)

i.   Propose two different mechanisms for the base-catalyzed hydrolysis of methyl acetate. Predict the labeling pattern for each mechanism, and explain how the isotopic labeling can be used to rigorously exclude one of your mechanisms.

9.   (continued)

ii.   Propose two different mechanisms for the hydrolysis of t-butyl acetate in acidic solution. Predict the labeling pattern and the products for each mechanism, and explain how the isotopic labeling can be used to rigorously exclude one of your mechanisms.

9.   (continued)

iii.   2-methylpropene is a significant product of the acid-catalyzed reaction. Explain how one of your mechanisms, but not the other, is consistent with this experimental observation.

iv.   Discuss the following reaction with your colleagues to identify important control experiments to test for trivial sources of the labeling patterns:

$$
\underset{\text{e.g.,}}{} \quad
\overset{\displaystyle O}{\underset{\displaystyle}{\underset{\displaystyle}{RC}}}\!OH
\quad \xrightarrow[\text{H}_2\text{O}^{18}]{\text{H}_3\text{O}^+} \quad
\overset{\displaystyle ^{18}O}{\underset{\displaystyle}{RC}}\!^{18}OH
$$

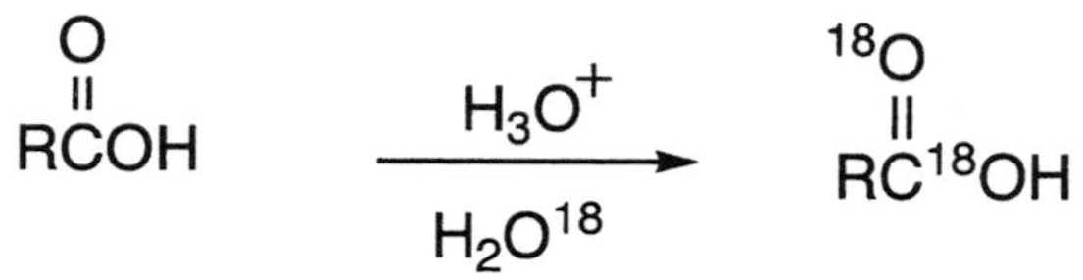

10.   a.   Consider a variant of the SN2 reaction in which the reaction proceeds with *retention* of configuration. Use molecular orbital theory to determine the feasibility of a front-side attack of a C-X bond by a nonbonding orbital of a nucleophile as shown in the following diagram:

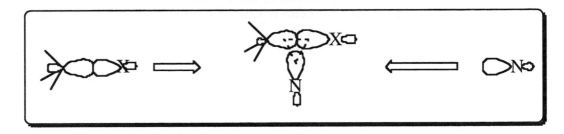

i.   Place the *relevant* molecular orbital(s) (MOs) of the C-X sigma bond on the left side of the paper and the orbital of the nucleophile on the right. Construct the molecular orbitals for the preceding interaction in the middle of paper. The nodal properties (i.e., symmetry) and the relative energies of all of the orbitals should be clearly illustrated.

ii.   Determine whether this interaction of a nucleophile and C-X bond is energetically favorable or not. What is your prediction about the feasibility of the reaction?

iii. Consider now the reaction between an electrophile and the C-X bond as described by your diagram. Determine whether this interaction is energetically favorable or not. What is your prediction about the feasibility of the reaction?

10.   (continued)

b.   Consider the cyclooctatetraene dication (i.e., cyclooctatetraene from which two *p*-electrons have been removed). (1) Provide a diagram of the energy levels of the pi molecular orbitals of the dication, showing how they are occupied by the pi electrons. (2) Predict whether the system will be aromatic or antiaromatic.

c.   Explain the differences in acidity:

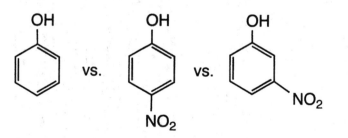

     OH   OH   OH

          vs.   vs. NO_2

                    NO_2

$pK_a = 9.9$ $pK_a = 7.2$ $pK_a = 8.3$

     +NH   +NH

              vs.

$pK_a = 7.0$ $pK_a = 2.5$

274

10.   (continued)

d.   Consider the dimerization of two substituted ethylenes, as shown in the diagram that follows. Use the molecular orbital representations and interactions to explain why the thermal cycloaddition is not observed, but the photochemical reaction is observed.

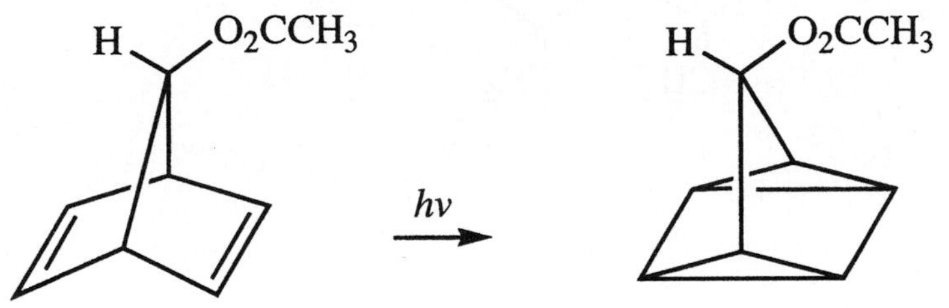

10.   (continued)

    e.   Propose products for the following reactions.

    i.

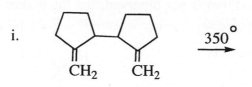 

    ii.

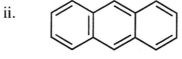

         anthracene   benzyne

    iii.

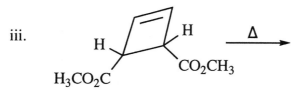

    iv.

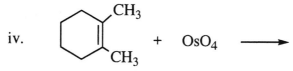

Give a careful representation of the 3-D structure of the osmate ester.

11.   a.   Specify the net ionic charge on the following peptides and the direction of electrophoretic mobility at pH 1, 7.3, and 12.

MetValPhe   LysAlaGly

AsnGluPro

b.   The products of deprotection by removing the BOC group are isobutene and CO_2. Propose a mechanism for this acid-catalyzed reaction.

$$\underset{\underset{O}{\underset{\|}{NHCOC(CH_3)_3}}}{RCHCO_2H} \xrightarrow{CF_3CO_2H} \underset{NH_2}{RCHCO_2H} \quad + CO_2 \quad + (CH_3)_2C{=}CH_2$$

11.   (continued)

c.   The tertiary (folding) structure is an essential element of the function of proteins. Consider the structures of the amino acids at physiological pH, and catalog different kinds of interactions that might be responsible for the tertiary structure. Give structural representations of the different interactions. Most of these will be noncovalent interactions, but the reaction of two cysteines to form a disulfide bridge is an important covalent interaction. Simultaneously, suggest reagents that might denature the protein by disrupting the interactions you have identified.

11.   (continued)

d.   In the presence of water, Cr(VI) oxidants convert primary alcohols to carboxylic acids via the general reaction

$$RCH_2OH \longrightarrow [RCHO] \longrightarrow RCO_2H$$

Pyridinium chlorochromate (PCC) is a Cr(VI) oxidant that is easily prepared in anhydrous form. As a result, PCC is the reagent of choice for converting alcohols to aldehydes.

i.   Recall the mechanism of oxidation of alcohols by Cr(VI) reagents, and explain how the presence or absence of water determines the outcome of the oxidations. (*Hint*: Recall also the reaction of aldehydes with water.)

ii.   Explain why the following compound reduces PCC in the absence of water:

$$\underset{(CH_3)_2C(CH_2)_3CH}{\overset{\underset{|}{OH} \quad \overset{O}{\underset{\|}{}}}{}}$$

11.   (continued)

   e.   Cu(citrate)⁻ reacts with glucose to give a precipitate of Cu_2O (Benedict's test). The appropriate half-cell reactions and the corresponding standard potentials at pH 7 are

$$2Cu(cit)^- + 2e^- + 2HO^- \rightleftharpoons Cu_2O + 2\,cit^{3-} + H_2O$$

$$\varepsilon^o = -0.09V$$

$$gluconate^- + 2H_2O + 2e^- \rightleftharpoons glucose + 3OH^-$$

$$\varepsilon^o = -0.44V$$

   i.   Write the cell reaction and calculate ε° for the cell. Does the equilibrium favor glucose or gluconate ion at pH 7?

   ii.   Use Le Chatelier's principle to predict how the cell potential varies as the pH increases.

   iii.   The pH of the Benedict's test solution is 12. Is the concentration of glucose at equilibrium higher at pH 12 or pH 7?

11.   (continued)

f.   Phospholipids are the major components of cell membranes. The fluidity of a cell membrane depends on the nature of the phospholipids. Membrane phospholipids from different parts of the leg of a reindeer have different compositions of fatty acids. Given that reindeer walk around in the snow, predict whether the proportion of unsaturated fatty acids is higher or lower closer to the hoof. Explain your reasoning.

12. We often make connections (i.e., learn) by recognizing analogies. Find the analogy in each of the following pairs of reactions:

a. 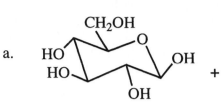 +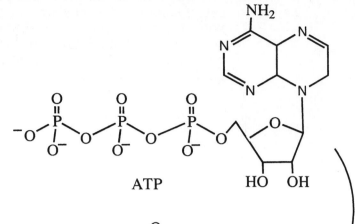

glucose   ATP

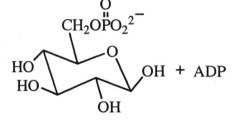

glucose-6-phosphate

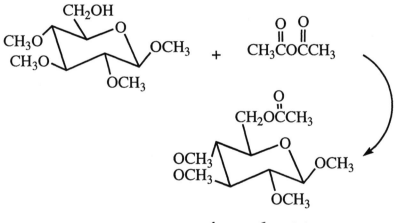

glucose-6-acetate

12.   (continued)

b.

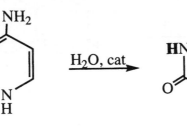

cytosine   uracil

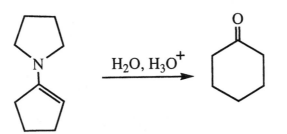

c.

CH$_2$OPO$_3$$^{2-}$
|
C=N$^+$H(CH$_2$)$_4$(CH$_2$)$_4$CHCO$_2^-$
| |
CHOH   N$^+$H$_3$
|
CHOH   →（:B，aldolase）→ BH$^+$
|
CHOH
|
CH$_2$OPO$_3$$^{2-}$

CH$_2$OPO$_3$$^{2-}$
|
C—NH(CH$_2$)$_4$(CH$_2$)$_4$CHCO$_2^-$
‖ |
CHOH   N$^+$H$_3$
+
HC=O
|
CHOH
|
CH$_2$OPO$_3$$^{2-}$

CH$_3$
|
C=O
|
CH$_2$ →（:B，aldolase）→ BH$^+$ + 2CH$_3$CCH$_3$ (O)
|
H$_3$CCOH
|
CH$_3$

12.   (continued)

d.

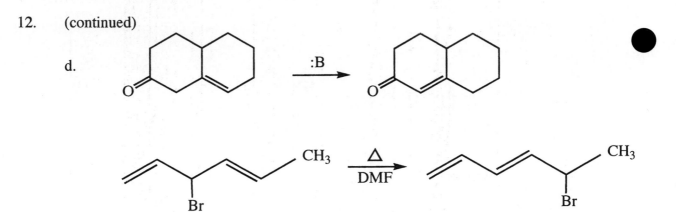

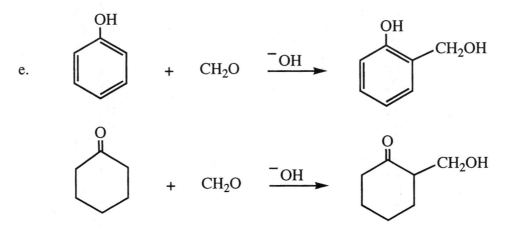

e.

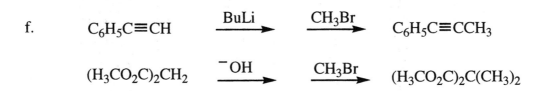

f. $C_6H_5C{\equiv}CH$ $\xrightarrow{\text{BuLi}}$ $\xrightarrow{\text{CH}_3\text{Br}}$ $C_6H_5C{\equiv}CCH_3$

$(H_3CO_2C)_2CH_2$ $\xrightarrow{^-\text{OH}}$ $\xrightarrow{\text{CH}_3\text{Br}}$ $(H_3CO_2C)_2C(CH_3)_2$

12.   (continued)

g.

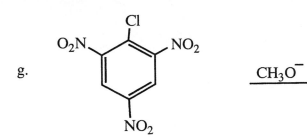 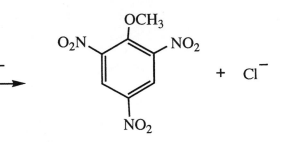

+ Cl^-

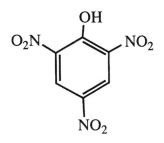 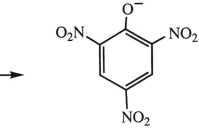

pKa = 0.3

h.

   +   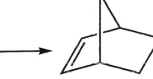   +

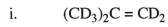   +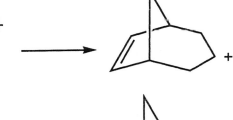

i. $(CD_3)_2C = CD_2$ $\xrightarrow{CF_3CO_2H}$ $(CH_3)_2 C = CH_2$

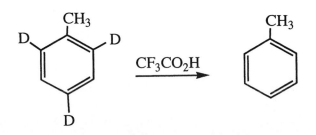

12. (continued)

j.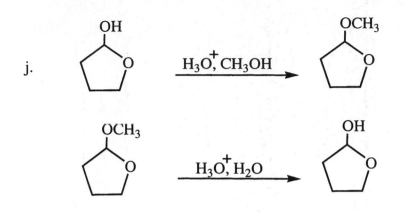

k.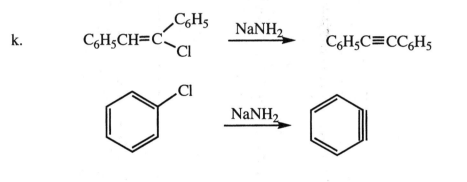

l. $C_6H_5CH_3$ $\xrightarrow[\text{CCl}_4]{\text{N-bromosuccinimide, } hv}$ $C_6H_5CH_2Br$

$CH_3CH = CH_2$ $\xrightarrow[\text{CCl}_4]{\text{N-bromosuccinimide, } hv}$ $BrCH_2CH = CH_2$

12.   (continued)

m.

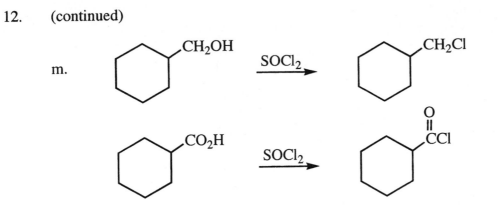

n. → z.   And the rest is up to you!   Work with your colleagues to find another 13 analogies.

**GOOD LUCK
ON THE
FINAL EXAM!**